What people are saying about

Creator and Creators

Many Thanks for your Great Wisdom and Artistry.
Victoria LePage, author of *Shambhala and Mysteries of the Bridechamber*

I would like to congratulate you on this great work. It is quite comprehensive.
Sanjay Raghav, MBBS, MD, DM, FRACP, adjunct senior lecturer Monash University, Department of Medicine, Therapeutic Director at Yarra Valley Living Centre, House of Gawler Cancer Foundation

...full of interesting ideas and concepts, this is a book which presents a possible view of life and existence. Much of it should be familiar to those with some experience of esoteric traditions and the authors' hypothesis that existence is a multi-layered hologram is a good addition to the esoteric literature.
Dr. Gregory Woodford, PhD in mathematics Melbourne University

Creator and Creators

Co-creation with Nature: A Synthesis of Spiritual Philosophy and Science

Creator and Creators

Co-creation with Nature: A Synthesis of Spiritual Philosophy and Science

Roza and Margarita Riaikkenen

Edited by Andrew Rooke

AXIS MUNDI BOOKS

Winchester, UK
Washington, USA

First published by Axis Mundi Books, 2018
Axis Mundi Books is an imprint of John Hunt Publishing Ltd., 3 East Street, Alresford, Hampshire SO24 9EE, UK
office1@jhpbooks.net
www.johnhuntpublishing.com
www.axismundi-books.com

For distributor details and how to order please visit the 'Ordering' section on our website.

ISBN: 978 1 78535 705 3
978 1 78535 708 4 (ebook)
Library of Congress Control Number: 2017937814

A CIP catalogue record for this book is available from the British Library.

Design: Stuart Davies

Printed and bound by CPI Group (UK) Ltd, Croydon, CR0 4YY, UK

Contents

Also by Roza and Margarita Riaikkenen

The Laws of Life

The Return of the Prodigal Son

To our Teachers

"Give me a place to stand on, and I can move the earth"
(Archimedes)

"Namely, the beauty of synthesis will remain for the whole of your life. Each investigator who devotes himself to even the least detail in the structure of the Universe arrives at it through the principle of breadth, and not through narrowness. Thus, cognition will be all-embracing. Verily, where burns the fire of knowledge, there has been ordained a luminous future"
(Agni Yoga, Brotherhood, 427)

Editor's Foreword

"As Above so Below" the Hermetic philosophers of ancient days would say, but what exactly does that mean for today's world? The book you have in your hands beautifully and succinctly speaks to a 21st century readership with a combination of Ancient Wisdom and the latest discoveries in modern science and mathematics. It presents the creation myths of old in language we can understand and appreciate. The wisdom of the ancient sages and how it can be practically applied to save our world permeates every page.

Following and developing themes from the authors' 2004 classic, *The Laws of Life*, this book combines mysticism and science to show how man relates to Universal Creation and his latent capacity as a co-creator in Nature's divine plan. Here you will find a broad vista of perspectives on cosmology and hints for the future based on cooperation rather than the exploitation of Nature. Drawing on the authors' scientific background and their own extensive life experience, combined with expert knowledge of Theosophy, *Agni Yoga*, and the modern Russian spiritual classic, *Kalagia*, this book presents explanations of the three worlds of Spirit, Subtlety, and Solidity and how they relate to each other. A huge range of related subjects are expertly presented including practical advice on why and how we should control our thoughts and how thought and consciousness work in a holographic Universe.

Ancient philosophic questions such as the power of Free Will versus Predetermination and the mysteries of how the One became the Many are explained in the light of ancient teaching and the latest scientific knowledge. Read on and you will discover practical applications of viewing the Universe as a manifestation of conscious energy, such as colour and other kinds of vibration. Brain science and reproductive medicine are seen in a new light.

Practical directions in life based on seeing ourselves as spiritual voyagers in a sea of energy are discussed, such as overcoming depression and anxiety without drugs.

Implications of looking at the world as a manifestation of light and vibration are presented in a clear and understandable language. Of special importance in today's world of social media, the book features a fascinating section on manipulators of our thought life with the use of modern technology. Perspectives on the mysteries of life after death are explained in easy to understand terms and with implications for daily life in today's high pressure world, for example, a discussion of suicide.

Technical discussion is combined with explanatory diagrams and parables to clarify the profound issues that absorb the thoughtful reader on every page. A glossary and comprehensive bibliography are provided for those who may wish to venture further on their own voyage of spiritual and scientific discovery.

So why not set your sail to the winds of change and venture into the unchartered waters of the Universal energy sea awaiting you in this fascinating new book from Roza and Margarita Riaikkenen!

Andrew Rooke.

Acknowledgements

We would like to express deep gratitude to our editor, Andrew Rooke, for his comprehensive and unwavering collaboration in our shared work with all the books, articles, and lectures that we have given and published over the past two decades.

Many thanks to mathematician, Dr. Gregory Woodford for his thorough review and thoughtful suggestions for improvement of the manuscript. We are grateful to neurologist, Dr. Sanjay Raghav for his appreciation, encouragement and valuable suggestions for the accuracy of the medical aspects of our book. Thanks to Professor Krishna Vaddadi for his thoughtful suggestions. Our thanks go to Paul Rooke for his expert advice over questions of economics and theosophy and for his thorough review of the manuscript.

Thanks to Richard Riaikkenen and Jasmine Brazilek who helped us along the way. Thanks also to Igor Paul Riaikkenen for his invaluable designer advice. Grateful appreciation to Rachel Kisin for her assistance with the introduction of the book, her encouragement and support. Thank you Alla Petrov for your trust in us and eagerness to help.

Our thanks to the Nicholas Roerich Museum (New York) and its Director Gvido Trepša, and to the Agni Yoga Society for the permission to cite quotations from the English translations of the books of *Agni Yoga.*

We deeply appreciate the encouragement and support of many others along the journey to create and publish this book. It is as much your voyage of discovery as it is ours.

Roza and Margarita Riaikkenen

Introduction

This book is on the one hand a sequel to the books, *The Laws of Life* and *The Return of the Prodigal Son* (published as a single volume in 2004) on the other hand it is an independent synthesis of spiritual philosophy and sciences.

While writing *Creator and Creators,* we aimed at synthesizing knowledge from various sources and combining it with our own experiences and realizations in order to discover mysteries of the universe. In our book, we describe universal laws and mechanisms as we understand them, crossing the 'great manmade divide' between the mysticism of ageless spiritual philosophy and modern science.

Such an approach gives us the advantage of being able to start from the very beginning of the universe and discover its reflection in any manifestation of life. It frees our hands from the limitations of purely materialistic science and enables limitless travel in the multi-dimensional environment of consciousness and energy. Along the way, we find answers to the enigmas and paradoxes which modern science faces in abundance.

The most important advantage of the synthesis of knowledge is that it allows the connection of all the concepts and mechanisms in a dynamic system. Every part of this system can be explained from the viewpoint of the whole; and the whole is a result of the interaction of its parts.

We are starting from the 'grassroots', from the very essence of the concepts of Consciousness, Thought, Spirit, Matter, Time and Space. We establish the language and explore the rules of how everything exists and is interconnected. We attempt to stay completely open to any new facts and information that allows us to take our next steps on the path of discovery.

Cosmic laws of which we became aware while writing our first book, *The Laws of Life* (2004), enable us to go further in the search for universal mechanisms of existence. We seek to find out how these mechanisms are working in science and in everyday life. We try to apply our new knowledge towards the illumination of Life's mystery.

There were times in human history when great thinkers used to make discoveries in natural sciences by simultaneously employing spiritual philosophy and scientific research. They were creating schools of thought that became beacons of knowledge, culture and education for their era.

Among these spiritual luminaries, the founders of the ancient Eastern schools of philosophy: Chinese Taoism, Confucianism and Buddhism (formed c. 500 BC); India's orthodox schools of Vedanta and Yoga (from approx. 400–450 BC), and non-orthodox schools of Buddhism; Arabic Sufism (from c.660) and Egyptian mystery schools that probably existed from the 1st millennium BC and possibly even earlier.

In Europe, schools were founded by a metaphysician, Aristotle (384–322 BC), a spiritual initiate, Pythagoras (born 571 BC), philosophers, Socrates and his student and follower, Plato, (c. 400 BC). In the 18th century, the discoveries of a student of spiritual philosophy, Sir Isaac Newton, laid the foundation for classical physics. But classical physics appeared to be only a small part of Newton's overall heritage of universal science. He was also a great mystic and his unpublished papers in the Bodleian library show he was primarily interested in Hermeticism and Alchemy.

These were schools of thought dedicated to the pursuit of knowledge, whatever it would be, from wherever it could be received, and merging it by means of philosophy. Sciences weren't separated from each other the way they are today. Together, they would provide the foundation for scientific progress, ethics and religious concepts, laws and rules of behaviour for nations and societies.

As it turned out, with the development of technologically

inclined societies, only materialistic knowledge has been accepted, applied, and recognized as 'science'. It served well for the development of the technical foundation of our modern civilization. However, when we start asking fundamental questions about consciousness, the origins of the universe, energy, matter, time and space, and even a definition of life itself, then we cannot come to reliable answers from the standpoint of modern science alone.

According to David Bohm, one of the most significant theoretical physicists of the 20th century, "Many thousands of years ago our culture was not broken into fragments as it is now. At that time science and spirituality were not separated. Since then they have grown far apart. In my view it is important to bring them together."[1]

Besides his other pioneering theories, Bohm contributed innovative and unorthodox ideas to the philosophy of mind [2, 3]. His ideas inspired neurophysiologist Karl Pribram, who developed the theory of the holographic nature of brain processes[4]. In collaboration with Denis Gabor, physicist and inventor of technical holography and mathematicians such as Roberto Llinas and others, Pribram came to remarkable conclusions we shall examine later in the book.

Recent academic research has begun looking into the previously forbidden field of spiritual sciences for assistance in finding explanations for some of the facts it encounters. For example, in his bestseller, *A Brief History of Time* (1988, reissued 2011), renowned theoretical physicist and cosmologist, Stephen Hawking considers some theistic models of the beginning of the universe together with those of modern science.

According to journalist and researcher, Lynne McTaggart, author of *The Field* (2008), world famous universities, such as Princeton and Harvard, nowadays invite healers of different denominations, monks, and shamans, to participate in their research into alternative mind/body therapies and their application in modern orthodox medicine.

Quantum physics finds itself in a particularly interesting situation. By going into the micro-world of atoms and subatomic particles, it came to the point of uncertainty: what are these particles in reality? Why do they sometimes 'entangle' and move together in something like a choreographed dance? Are they really particles? Are they vibrating strings? Or are they energy waves?

In January 1998, Edward Witten, an eminent professor of physics at the Institute for Advanced Study in Princeton, recipient of the 1990 Fields Medal, mathematics' highest prize, member of the National Academy of Sciences, recipient of a MacArthur Prize, the Dirac Medal, and many other honours, said in his public lecture *Duality, Spacetime and Quantum Mechanics*[5], "In the last few years, physicists have learned that the different string theories discovered and studied in different ways are limiting cases of a single, more powerful theory, known as M theory. 'M' stands for magic, mystery, or matrix, according to taste."

'Matrix' or mathematical description, in string theory, replaced the unpredictable quantum particles with hypothetic 'strings' which produce vibrations. However, Witten also said, "You can describe much of physics by particles or waves, but the waves are more fundamental; only the less fundamental particle view has been generalized in the string theory."

As we see it, mathematics doesn't describe literally 'strings', but rather waves, vibrations, produced by the strings. Waves or vibrations can originate from any energy source. They are in their essence rhythmically moving energy. According to Witten, waves or vibrations, i.e. rhythmically moving energy, constitute the fundamental substance of the universe.

In this book, we further develop the idea of rhythmically moving energy as the mother-substance from which everything in the universe emerged. Like Bohm and Pribram, we see energy (the energy of Light) manifesting the physical universe through the mechanism of holography. However, for the full picture of manifestation, we need to know the source of this energy, and also the source of the template for manifestation. So, where is this creating force which is initiating, forming and defining processes in the universe?

We cannot discover this force whilst remaining within the paradigm of our habitual materialistic 'particle based' science. The deeper science delves into the elements from which matter derives, and the smaller the particles, the more unpredictable their properties and behaviour.

In quantum physics, as soon as we 'encounter' the smallest particles, 'leptons', we become stuck in uncertainties and probabilities. Actually, quantum physics is based on the uncertainty principle.

In their book *Quantum Enigma (2011),* physicists Bruce Rosenblum and Fred Kuttner cite the renowned Werner Heisenberg: "In the experiments about atomic events we have to do with things and facts, the phenomena that are just as real as any phenomena in daily life. But the atoms or elementary particles themselves are not real; they form a world of potentialities or possibilities rather than one of things or facts."[6]

The authors of *Quantum Enigma* also state: "Quantum theory tells us that any look, anything that provides information, collapses the previously existing state" [7], which implies the significance of a conscious observer in the quantum events. From these findings the authors of *Quantum Enigma* come to the enigma of consciousness and its role in the micro-world of quantum physics and hence also in the macro-world, which consists of the micro-world content: "Starting with quantum mechanics we have encountered consciousness"[8].

Consciousness has become one of the greatest mysteries of science. Where is its origin? In the brain or beyond? Aren't different levels of consciousness responsible for the success of our experiments and scientific conclusions? In our book, we are dealing with this enigma of physics by starting from consciousness and exploring the ways and mechanisms of how it is creating the world that we know.

Putting the findings of experimental sciences into the context of spiritual philosophy – metaphysics and its certain and exact cosmic laws, enables a broader outlook and understanding of their knowledge. On the other hand, the discoveries and models of modern science enrich philosophy with new content and mechanisms and raise some fresh questions for us to ponder.

As spiritual philosophy stands, we consider Consciousness to be at the beginning of creation as its main creative force. We consider Matter to be a product of the workings of Consciousness.

Despite the scepticism of modern science, spiritual philosophy has never ceased to exist and has continued to develop throughout the centuries. Amongst many other sources of information, we refer mostly to three books of spiritual philosophy: *The Secret Doctrine* by the principle founder of the Theosophical Society, H.P.Blavatsky (1831–1891), Helena Roerich's *Agni Yoga* series (transmitted by her Master, starting from 1920 up to 1941, see Appendix 1) and *Kalagia* (1993) – a book of metaphysics dictated to Alexander Naumkin by his spiritual Teacher in the Altai Mountains in Russia (see appendix 2). *Kalagia* has been published only in Russian since 1993. It has never been translated into English apart from some quotations that you can find in this book and in our other book, *The Laws of Life* (2004).

We are attempting to synthesize spiritual philosophy and modern science. Science is based on experimentation and observation of the phenomena of the world as our senses and tools perceive it. We apply the multi-dimensional and multi-aspect understanding of spiritual philosophy in order to explain the observed phenomena of the world. The whole explains the part, and we are seeing science as a part of a bigger universal picture where Spirit, the Creator, and we, Its co-creators, are naturally collaborating in the creation of the beauty of life. Synthesis also allows us to explain the bigger picture of the universe through examples provided by our personal experiences and the experiences of other researchers.

From this picture, the purpose and meaning of life can be derived for everyone. We found, as have our predecessors in spiritual research, that such understanding may also be of infinite use for finding solutions for our own daily life problems.

We composed the book out of the knowledge that we have received from diverse sources, including our meditations and the discoveries and abilities that we achieved throughout our lives of experience and learning in the fields of science and spiritual philosophy alike. We attempt to share our methods and understanding with the reader. An attentive reader will learn how to look and see behind and beyond the visible manifestations of life situations that we commonly accept as our material reality. We hope that such an approach will help him or her to solve many of the tasks and problems of their life from the standpoint of their newly gained understanding.

Perhaps the main discovery for us was that perceptible, almost palpable awareness of connectedness of everything in universal consciousness. We try to prove it by showing how the same laws and mechanisms are working in different manifestations of life: the universe, nature, our mind, and our bodies.

We don't reject any source of knowledge. This is in accordance with our principle which we proclaimed in *The Laws of Life*: "if any rule is working in the macro-world as well as in the microworld, in all their manifestations – then it is a law of life".

So, let's start from the very beginning!

Chapter 1

The Process of Manifestation – Creation

The Urge to Live

Is there any particular reason for our existence? If yes, how did everything we know and see come to be? Humanity has been asking these questions as long as it has existed. On the journey to find answers to these questions, we make discoveries that lead humanity forward. Every time the consciousness of humanity takes a step forward, we see the picture of our universe anew.

The Introduction to Creation (Manifestation): Let us start from an axiom – the universe exists eternally; periodically changing the form of its existence. The fundamental property of the universe is its pulse, its rhythm or the 'heartbeat of the universe'. All life is rhythmical: the motion of cosmic bodies, heartbeats and breathing of living essences, cycles of nature, movements of waves.

When you exhale to the end of your capacity, do you experience an urge for inhaling again? And vice versa, when you inhale, do you have an urge to exhale? As *Agni Yoga* puts it: "In Cosmos lives that conscious power which is called the cosmic rhythm and the whole of human life depends on the cycle of this rhythm."[1]

The universe is 'breathing' in accordance with its rhythm – manifesting as a physical entity and returning to the state of No-Thing – pure energy: breathing out (while manifesting) and in (to return). Hindu philosophy depicts the two conditions of the universe as 'Manvantara' (the manifest condition) and 'Pralaya' (unmanifest condition). In Hindu understanding, Brahma is exhaling the universe and inhaling it.

Theosophy puts Brahma as the principle creating the manifest universe and returning with it into No-Thing. One of the fundamental assertions of the *Secret Doctrine* is "the absolute universality of that law of periodicity, of flux and reflux, ebb and flow, which physical science has observed and recorded in all departments of nature. An alternation such as that of Day and Night, Life and Death, Sleeping and Waking, is a fact so common, so perfectly universal and without exception that it is easy to comprehend that in it we see one of the absolutely fundamental laws of the universe."[2] Whatever the pair of opposites, the idea remains that the manifest universe appears and disappears in accordance with its rhythm.

When we are discussing the subject of appearance of the universe, we obviously mean its 'Manvantara', manifestation from No-Thing. In the very depth of any material thing, as if looking through a high definition microscope, we eventually see No-Thing – the Source of Everything. So, let's start from No-Thing, or the 'Great Void' – the Source of our Universe.

If nothing else exists, No-Thing is simultaneously Every-Thing. People may feel and call this Nothing/Everything – God the Almighty. And almighty Nothing/Everything is, being the Source of Everything and having Everything as its potential. We will call this Nothing/Everything – Spirit. We accept that Spirit is unlimited and omnipresent.

H.Blavatsky writes in *The Secret Doctrine* that the 'Divine Thought' is the metaphysical cause of the existence of everything, and the desire to live is its reflection: "This desire for a sentient life shows itself in everything, from an atom to a sun, and is a reflection of the Divine Thought propelled into objective existence, into a law that the Universe should exist."[3] Karl Jung mentions the 'urge to live' and calls it a feminine aspect. We can also understand and accept this 'urge to live', or 'the desire for a sentient life', as Spirit's feminine aspect.

The Fire of Consciousness: In the Nothing/Everything Spirit already has the Fire of Consciousness as Its potential. To live, for Spirit, means to create, and to create Spirit has to make this potential work. Thus starts the work of the creating trinity. A creating trinity is based on the interaction of the Seeker (the one who sets the aim), the Sought for (the aim), and the Search as the process. The Trinitarian mechanism reveals itself in different situations and acting trinities. We will discuss in detail the working of a Trinity as the main mechanism of creation later in the book (see "Trinity – a Mechanism of Creation").

Kalagia (see appendix 2) describes the mechanism which puts the Fire of Consciousness into action through the working of the Trinity of Spirit, Fire of Consciousness and Motion from them and in them. It means that Spirit becomes the Seeker, Fire of Consciousness becomes Spirit's Sought for, and the Search, the process, is Motion from them and in them. Motion, like a current of electricity between plus and minus (and actually the very source of electricity) turns Fire of Consciousness into the flow of primeval energy (the Tibetan 'Fohat'). We can also say that it starts the flow of energy of Divine Thought – the creative power of life. In this way, Spirit becomes the Creator, and His feminine aspect, which reflects the Divine Thought, becomes the energy of creation (the Hindu goddess 'Shakti').

A Ray of Light emanates from Spirit – an aspect of Spirit's Fire of Consciousness (the intensity of Fire, according to *Kalagia*). The Creator needs a tool, a 'building material' for the creation and Light becomes this tool. Also, we can see this Ray of Light as being represented by many Rays existing in a multitude of dimensions. This Ray (or Rays) is the instrument of the holographic process of manifestation.

The Holographic Process of Manifestation: The meaning of the process of creation (manifestation) is that the information of Spirit's Thought gets turned into a holographic image, a hologram which would be animated by Spirit enabling it to live.

As physics knows and applies, information can be stored by an interference pattern of waves of light. The technology of holography (see fig.1) is based on the interference of two beams of light. One of them is a coherent beam, usually a monochromatic single wave laser beam. This beam is split by a 'beam splitter' into two beams. One of the beams is the initial 'reference ray' which provides a basis for the formation of the interference pattern. In technology the reference ray is usually diffused by a lens and directed to a photographic film or plate.

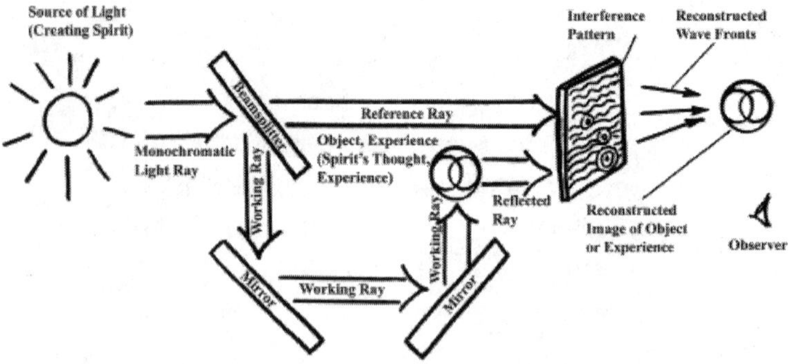

Fig. 1.The Holographic Process of Creation (Manifestation)

The other beam passes through an object or experience and becomes the 'working ray'. It is split from the initial beam and directed to the object by a system of mirrors and a diffusing lens. Meeting, illuminating and interacting with the object, the 'working ray' receives an experience. After being reflected from the object, the 'working ray' also falls on the photographic plate.

Interference of the 'reference' and 'working' rays produces an interference pattern on the plate. This pattern contains information about the object (or the beam's experience) which is based on the phase difference between the reference and working rays. The pattern becomes a potential for the hologram's manifestation.

The pattern on a plate can be later illuminated and read as an image by a coherent 'reconstruction ray'. In technology, as shown in fig.1, it can be the same laser light used for making the interference pattern, i.e. the 'reference ray'. Then an observer is able to see a three-dimensional image of the object or experience – the hologram.

In the hologram, any part reflects the whole. We can cut the plate with the interference pattern into pieces and every piece will reproduce the whole image of an object or experience. The final image of the hologram, as it is seen by the observer, depends also on the observer's judgement. The observer's memory has a picture of what was before as a baseline for comparison with the hologram that the senses bring to the observer's attention. So, whatever the observer perceives, it is the difference between the original picture and the observed changes in it.

Spirit manifests the universe through a similar process using Its own Ray of Light as a 'reference ray' (see fig.1). This initial Ray becomes also a 'working ray' after any encounter with an experience. Spirit's Thought is an experience for Spirit. The 'working ray', in this case, containing all Spirit's Consciousness, yet carrying the imprint of the experience of Spirit's Thought. In the uninterrupted process of Creation, the reference ray serves simultaneously as a reconstruction ray.

In this way, a hologram of Life is produced and recorded on the 'film' of Spirit's inner Space which *Kalagia* calls Spirit's Memory. According to *Kalagia,* Spirit's Memory is also Spirit's Body – Soul or information-energy field that becomes the reality of 'Prima Materia'.

Spirit as the first Observer observes Itself through experiencing Itself and reflecting Itself in the hologram of Its experiences. In this way Spirit produces the first hologram – the Logos. This hologram exists in the Spirit's Consciousness, in Its Memory.

Spirit imparts Its Consciousness to any entity which It

manifests be it a mineral, vegetable, animal, or a human being, making all of them parts and simultaneously active participants in the holography of manifestation. As participants, all living beings become observers and creators of the hologram of life, allowing Spirit to observe through their 'eyes'. As parts of the holographic picture of creation, they reflect the properties of the whole. We can also find multiple holographic mechanisms in their parts, such as genetic apparatus or the central nervous system and the brain. We will further discuss some of these in the book.

An observer perceives a hologram, compares it to the picture already existing in memory, and adjusts his/her thought accordingly. This adjusted thought becomes the experience of the 'working ray' for the creation of a new hologram in the unending process of manifestation of the multifaceted hologram of life. This process, for example, is reflected in sunflowers turning their heads to the Sun and a mole digging into its burrow.

Appearance of Time and Space: We have to have in mind that there is no sequential Time or outer Space at the starting point of manifestation. The Reality of Spirit doesn't need such Time or Space until it starts manifesting the holographic 'Illusion' of Matter.

Kalagia says that no Matter could continuously exist in the timeless Space of Spirit – it would immediately explode and turn back into Fire. So, Time became an attribute of the holographic process and the means of stabilizing and carrying the holograms of Matter.

The cosmic rhythm contains a potential for Time to take a form of a wave, a spiral 'serpent', as it becomes the inner space of Matter. As *Kalagia* defines: "The inner space of Matter is Time, the substance Time or the hologram of the past."[4] This grand Illusion provides the environment for life, and life, reflecting the cosmic heartbeat, produces its variations of rhythms of the manifest universe.

Time is interconnected with Matter, and so is Space. We cannot say what is manifest earlier, and what later – Time or Matter, Space or Matter? Everything in an instant? *Kalagia* also states that "Time and Space are closely interconnected. As the substance of Time (the hologram) is the inner part of Matter, the Plasma of its ingredients, so Space is 'the outside' of Matter, Plasma outwardly inherent in the ingredients of Matter. They are united by Fire."[4]

We discussed the subject of cosmic Time in *The Laws of Life* and will continue discussing its aspects in this book. "Time, besides its flow, possesses also its density. The flow is measured by a clock, density – by a galvanometer."[5] Any changes in the order of the physical environment change Time density, and vice versa. The changes of Time density immediately spread all over the Space and change the order of the physical environment. So, when the density of Time changes, then the whole flow of events changes too. We can also say that Time can be 'squeezed' and 'stretched' like a spring because Time is a spiral, wave, or 'serpent' as it is called in some ancient cultures and occult texts.

These are the properties of Time anywhere in the Cosmos. When observing from the Earth, the cosmic Time behaves in conformity with Einstein's general theory of relativity. It showcases its non-linear 'curved' nature, and it slows down in the vicinity of massive solid manifested material bodies such as the Earth.

The Role of Chaos: Spirit and its Fire of Consciousness invisibly maintain their presence within each and every manifestation of Matter, filling it with life. No Spirit presence – no life.

To live, any system should develop. We can see that Matter, while sinking into its illusion and producing its 'solid forms', deviates from the perfect order of the spiritual potential, producing elements of chaos. The Solid World, at the end of its complete materialization, comes to the most pronounced state of chaos. However, according to *Kalagia,* chaos is also a necessary feature for the existence of the world:

> *...Chaos is a connecting chain in the structure of a system – it secures the wholeness, integrity of the system, but simultaneously is a factor of destruction. Therefore, the systems apply their fiery powers to eliminate Chaos, i.e. systems, rejecting Chaos, are processing it in themselves and using as fuel to maintain their energies.*[6]

This is the mechanism of stabilizing and maintaining the world in changing circumstances. On the one hand, to dispose of all the elements that are already not supported by Spirit and therefore not living anymore. On the other hand, to 'recycle' these elements into something suitable and useful for the system of life, and in this way to maintain the flow of energy.

Entropy is an inherent property of Matter and a measure of movement from Order to Chaos. However, living systems possess the ability of turning Chaos into Order.

According to *Kalagia,* "Life is the endless transmutation of death into the growth of Spirit, into the concentration of Its possibilities. This is an aspect of Matter's spirituality – its aspiration to the Future, and, in this way, the formation of the ingredients of Order from Chaos."[7] We human beings are doing this consciously, by going through the experiences of our lifetimes and coming to awareness as a result of our experiences. We are burning Chaos in our inner fire and using our 'spiritual crucible' to develop and to produce new Order.

Universes Manifest and Unmanifest. Dark Energy: The process of the universe's manifestation – the unfolding of a Manvantara – is happening continuously by the rays of Light manifesting Matter in Time. Still more potential universes are in a state of

Pralaya – rest – in 'no Time' (or completely 'squeezed' Time). Their Light doesn't spread, and constitutes that which we know as 'dark energy'. This is the potential for the future universes and worlds. When their Manvantara comes into being, Time and Space appear and Light spreads and becomes visible. Each of them will be seen where, for now, there is only dark energy – universes in waiting.

Trinity – a Mechanism of Creation

The Enigma of Trinity: From ancient times until now, we can see the images and signs of trinities in the artefacts of almost every religion and spiritual teaching.

In 1939, the famous spiritual traveller and painter Nicholas Roerich published an article *Banner of Peace*, where he described the sign of Triunity (three dots in a circle, see fig.2). This sign can be found in different parts of the world, and especially often in the Tibet and Himalayan enclave where it is understood as the sign of Shambhala. Roerich noted that this sign "appeared to spread all over the world" in India and Tibet, in the Sky Temple of China, and also in the images of Jesus Christ and Madonna in Europe, on the shields of the Crusaders and armorial bearings of the Knights Templars, and in the paintings of the artists of Spain and Italy. It can be found as far back as in the ornamentation of pottery of Neolithic times.

We could find examples of some trinities representing gods in different religious systems around the world: the gods of Greece (Zeus, Athena, Apollo), of Egypt (Osiris, Isis, Horus), of Rome (Jupiter, Venus, Mars), of India (Brahma, Siva, Vishnu), of Christianity (Father, Holy Spirit, Son), and the three columns of the 'Tree of Life' in the Qabalah of Judaism. Roerich stated that the sign of the Trinity was probably the most wide spread and omnipresent symbol of all times.

What does it mean? Is it just a coincidence or rather a sign of a law, an active mechanism in the universe?

The Trinities of Creation: H.P.Blavatsky researched the subject of trinities in depth in her books *Isis Unveiled* and *The Secret Doctrine*. She mentions the three trinities that "emanate and are confounded" in the "Being existing through and of itself"[8]: the initial, manifested and creative triads. The initial triad is defined as "ever existing" and "the most metaphysical" one.[9] Qabalah also vaguely mentions the initial trinity as 'veils' beyond human comprehension, while naming its components: 'Ein' (the Nonexistent or Hidden), 'Ein Sof' (the Hidden Unlimited) and 'Ein Sof Aur' (the Hidden Unlimited Light).[10]

Fig.2. Shambhala's Sign of Triunity

The second Blavatsky's triad – the 'manifested' trinity of Agni, Vayu and Surya – can be understood as the 'reducing transformer' transforming the highest vibrations of the multi-dimensional Fiery World (Agni) into the vibrations of the less-dimensional manifested Stars, such as our Sun (Surya). Stars that become the source of life for the manifested stellar systems everywhere in the universe.

The third Blavatsky's triad – the creative one – is made of Brahma-Creator, Siva-Destroyer and Vishnu-Protector. Their creative process is complete as it includes the functions of creation of the world, its development meaning destruction of everything obsolete and formation of new life, and protection of the new offspring of life till its maturation.

Another aspect of a working trinity is reflected in Qabalah, where after the initial trinity of 'veils' covering the most mystical concepts of the teaching, the next trinity is Keter, Hokmah and Binah. In her book *The Mystical Qabalah,* one of the most respected experts on Qabalah, Dion Fortune, refers to Blavatsky's idea about the impossibility of manifestation without the division of the One into the Pairs of Opposites which creates a trinity. So, as Keter is 'the Crown', the highest level of consciousness, it divides into Hokmah, which represents 'the Highest Father' – the masculine principle, and Binah – 'the Highest Mother' – the feminine principle.

The Secret Doctrine meanwhile depicts another situation where a trinity participates in the process of manifestation by gradually lowering the frequency of vibrations: "The One (Deity) becomes Two (Deva or Angel) and Two becomes Three (or man)".[11] These essences represent three forms of the United World: the Fiery World of the Deity, the Fine World of the Devas and the Solid World of Humans. We can also add that in the human being the physical body belongs to the Solid World, the Soul to the Fine World, and the Spirit to the Fiery World (see further "The Three Worlds of Existence").

The Trinity that is called in Christianity 'the Father', 'the Son' and 'the Holy Ghost' can be looked at also as a Trinity of 'the Father', 'the Son' and 'the Mother'. If we are looking at the process of creation (manifestation) through this Trinity, we can understand 'the Father' as the creating Spirit, 'the Son' as 'Logos' – the first manifestation of Spirit, and 'the Holy Ghost', or 'the Mother', as the divine Love energy (Shakti).

Trinity's Mechanism: As we already discussed in "The Urge to Live", we call the three components of a Trinity: 'the Seeker' (the one who sets himself the aim), 'the Search' (the process and its tools) and 'the Sought for' (the aim). According to *Kalagia,* the main objective of the Trinitarian mechanism that puts into action the primeval world of Spirit is Beauty:

> *Spirit is the Seeker; Fire is for the Highest the Sought for; and motion from them and in them is the Search. Their collaboration one in another is their Co-Search, and Matter is the Sign of the symbolic image of the Product of Co-Search. We are yet unaware of the Product of Co-Search, though we call it Beauty.* [12]

The depicted Trinity shows us how Spirit starts the process of creation from Its condition of 'Nothing/ Everything'. Spirit, the 'Seeker', begins the process of Motion (the 'Search') in Its search for the creative Fire, the 'Sought for' (the 'Fire of Consciousness' as we spoke of it in "The Urge to Live"). When the initial Trinity is put into action, then the Co-Search of Creation begins. The final product of the process that Spirit started is Beauty.

Thus we can see Spirit as the 'Seeker' when we look at the very start, the initiation of the process of Manifestation. In the process, the roles can change, and the participants of the process may appear in different roles in different trinities. For example, Spirit becomes the 'Sought for' for a human 'Seeker' on a spiritual path of evolution; and for this evolution the 'Seeker' has to find and put into action his or her inner Fire, which will play the role of the 'Seeker's' 'Search'.

Kalagia tells us that Spirit creates worlds by Fire and Light, and these worlds initially appear on our plane of reality as holograms of Light. H.P.Blavatsky also writes in *The Secret Doctrine* that the human body appeared initially as a "luminous incorporeal form"[13] and gradually became solidified in its physical material.

The trinities are playing the role of creation mechanisms – engines which propel the endless stream of Life into existence. In our everyday life, we also apply the Trinitarian mechanism in any process of creation. When a builder starts building a house, he becomes the 'Seeker'. The house is his 'Sought for'. His work with his skills and tools is his 'Search'. When the job is finished and the house is built, the house itself becomes the "sign of the symbolic image of the product of co-search". The real product of the builder's work is Beauty which also includes the new skills and understanding acquired by the builder.

The Birth of Archetypes

How Spirit is Shaping Matter by Thought: The objective of holography is to manifest the 'experience' passed by the 'working ray' (see fig. 1). The 'urge to live' evokes Spirit's Thought, which, for Spirit, is an experience. Spirit's Thought-form becomes recorded on the 'film' of Spirit's Memory as an interference pattern of the 'reference' and 'working' rays. This is the potential for the hologram of Life. Then the Spirit's 'reconstruction ray' manifests a hologram of Spirit's Thought-form.

Spirit's Thought-form, the creation of Its Consciousness, is of highest order and harmony, and we see the elements of this order in the laws and mechanisms of life as they unfold to us. The shapes of holograms manifested out of the pure content of the experience of Spirit's Thought reflect the perfect order of Spirit's Consciousness. These shapes become archetypes for manifestation of life by all the observers, and simultaneously, participants of the grandiose 'play of life'.

Let's examine what could the Spirit's experience be like, how does the Spirit's Thought produce the archetypes, and what does the resultant matrix of archetypes represent.

Division of the One and the 'Cosmic Magnet' of Unification: Spirit's experiences start with Its awakening. Spirit is Infinity in its Wholeness, in all of Its dimensions, which we can symbolically express with a circle (Qabalah calls it 'The Hidden Unlimited').

The essence, the potential of Spirit, remains in every part and particle of life created by Spirit. It remains as the 'core', or the 'seed', from which every living thing is originating and growing. This 'seed' could be symbolised by a dot in the circle (see fig.3a).

Spirit is in a condition of complete Unity. Life is born in and from this Unity and is manifested holographically. That's why every part and particle reflects the whole. The 'seeds' of creation in all the seemingly separate beings are unified in their essence. Hence the power of attraction arises. The forces of attraction in each essence, from a sun to an atom, take their beginning in their core. *Agni Yoga* calls this attraction of the parts of the unity – the 'Cosmic Magnet'. The Spirit's 'Cosmic Magnet' becomes the real

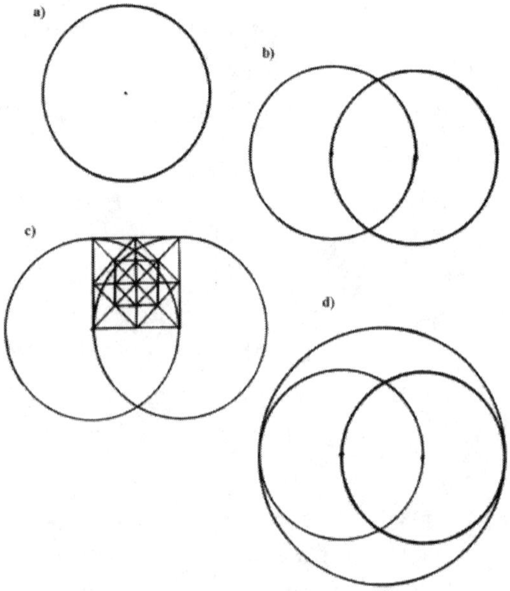

Fig.3. First Archetypes of Creation
a) The One, the 'Seed' of Creation
b) Division of the One into Two Opposites
c) One of the "First Experiences of Void's Crystallization and Formation of Space Substance" (Kalagia)
d) The 'Cosmic Egg'

source of any attraction, including gravitation, which is perhaps the most familiar aspect of this principle to us.

The state of unity is a static condition. But the urge to live and the vibration of the initial rhythm both create a 'swinging pendulum' that leads to the division of the One. This division could be graphically represented by a circle divided into two opposites, two intersecting identical circles, as shown in fig.3b.

We can say that division is Spirit's conscious sacrifice, where Spirit sacrifices Its wholeness to enable the manifestation of life. Division into two circles also represents the process of separation into any opposites: Light and Shadow, Feminine and Masculine, etc. Spirit and two opposites together make a working trinity, starting the process of manifestation. The division incurs polar tension – repulsion of division and attraction of 'cosmic magnetism' simultaneously.

The poles are repelling each other, but the 'Magnet' of Unity is always there, as is the yearning to return, to unite. It is a shift which, on the one hand disrupts the unity, but on another hand requires and makes possible further continuous motion and creation of something new. As *Agni Yoga* says:

> *The correlation of cosmic transformations is called cosmic creativity. When Cosmos shifts the forces, the balance of the spheres is disturbed. When the balance of these forces is upset, forces in the space are drawn into a new tension. Thus, when Cosmos shifts, all spheres are shaken. Indeed, all forces expand in response to an attraction, and the cosmic harmony is intensified by the Cosmic Magnet. Thus is Infinity created.*[14]

'Weaving' the Archetypal Shapes and Mathematics: The opposites are swirling in opposite directions producing whirlwinds of energy (electromagnetic vibrations) which follow in their motion the harmony (order) of Spirit's Thought. The pattern of this motion in the electromagnetic field can be likened to weaving or to making a spider web, where all connecting lines actually are 'lines of force' of the electromagnetic field. The ancient Egyptians expressed their understanding of this process as 'the Goddess Neith weaving the world with her shuttle'.[15]

The result is the matrix of archetypes which lays the foundation for all the shapes of created Matter. The symbols shown in fig.3 are some of the first archetypes as we human observers see and interpret. The symbols may well be the observer's conjecture. But, because "The Creator is inseparable from the Creation, the Creation is inseparable from the Creator" [16], so our thought is taking these images from the Creator's Thought.

Nothing is accidental and the pattern of the archetypes is built according to the harmony of Creation. Spirit's Thought forms the order of Creation and it appears that humans from ancient times onwards are seeing this order through number and pattern. We call this order – 'mathematics'. The order of the pattern we call – 'geometry'.

Mathematics is the way of orderly division of the One. It defines the order of division. Geometry defines the pattern of division. This provides Creation with the variety of forms. Not accidental, but orderly forms. The orderly repetition of forms, so-called 'fractal geometry', as in the processes of crystal growth or galaxy formation, shows the main property of the hologram – the whole reflected in every part. We can see the beauty of such geometry in the complex shapes of flowers and other objects of nature.

The Archetypes: The archetypal pattern that we can see as the expression of Spirit's division is the 'Cosmic Egg' representing the trinity of two opposites in unity within the circle (fig.3d) or the egg. Ancient myths and religions direct us to the primeval opposites as to the 'Great Father ('Divine Father') and the 'Great Mother' ('Divine Mother'). The 'Cosmic Egg', then, is an image of the 'womb' where the archetypes of Matter are born. (The

Hindu 'Hiranyagarbha' – the 'golden womb' originating from the primeval 'seed').

It may occur to us that the simplest kind of an archetype would be a straight line. However, most of the time a straight line is only an approximation showing the direction of motion or it is how our brains are used to interpreting an advancing wave. The prevalent archetype in the universe is a spiral form corresponding with the vibration of the cosmic rhythm, the 'serpent' of Time, and with the movement of the waves of Light. We can see a multitude of spiral forms in all sorts of environs, for example, spiral galaxies in the cosmos, cyclones and tornados in the air, whirlpools in water, and shells, fingerprints and much more in solid matter.

These are examples of the first archetypes as we see them. The archetypes created in the all-dimensional Spirit's Memory by the unlimited Spirit's Thought can be orderly shapes of different levels of complexity.

In our three-dimensional world of matter, the maximum of dimensions in which we see material forms is three. So, for example, Plato saw the first archetypal expressions of three-dimensional nature in his so-called 'Platonic solids' (see fig.4). Mathematics calls them 'regular polyhedra', which means 'many bases', and proves that they represent the only five types of three-dimensional forms equal in all directions. Plato identified four of them with states of matter (fire, air, water and earth), and the fifth – with cosmos ('Heaven above').

Every observer – a particle of Life in which Spirit abides – perceives and interprets the archetypal holograms in accordance with the observer's state of mind, which in turn produces experience for the 'working ray' and consequently a new hologram. The holographic picture of life as we see it is being continuously manifested out of experiences of all the observers.

All living beings carry the images of the archetypes of the Perfect Picture of Creation, the crystals of Spirit's Thought,

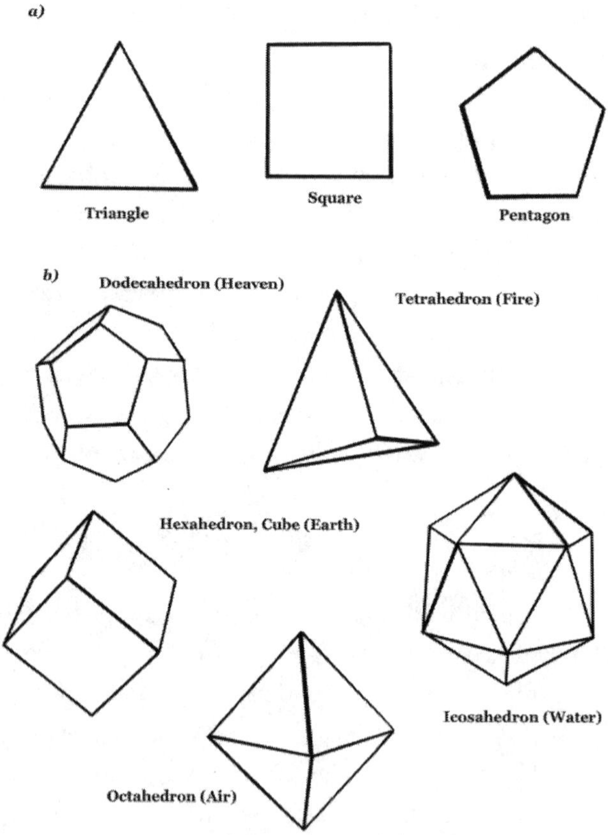

Fig.4. Regular Polyhedra – Platonic Solids
a) Bases of Regular Polyhedra
b) Regular Polyhedra

within their consciousness. Such crystals become magnets for the human thought, and through this they are moving humanity's thought forward. Not all of us can yet consciously tap into the depth of Spirit for a pure and whole archetypal pattern, the crystal of Spirit's Thought. Those mature Souls who went through the wholeness of experience and spiritual maturation in many incarnations, the great thinkers and sages of all times, the artists, composers and philosophers, the scientists making 'the cutting edge' discoveries – it is they who 'dig up' these jewels for the benefit of humanity.

Our human consciousness transforms these archetypes into forms that conform to our human understanding and reflect our common visual perception. That's why we can see similar themes and images in different cultures, in their myths, religions and fairy tales: the Father, the Great Mother, the Great Serpent, the Egg, etc. We are emulating these forms and through many lifetimes of experiences and learning are building ourselves into new forms that will eventually become archetypes. Like Christ or Buddha.

The Three Worlds of Existence

The Three Worlds of Existence: The Solid World imagery, though very familiar to us, neither covers all the dimensions of consciousness nor contains all the vibration frequencies of existence. Materialization of archetypes begins initially in the other two worlds – the Fiery and the Fine (Subtle) ones. These two Worlds are representing more dimensions and vibrations of higher frequencies than the Solid World.

All the holograms of Matter are contained as the Illusion of the Past in the Memory of Spirit, in the Space of Spirit's Consciousness. The three types of forms of Matter belonging to these three Worlds correspond to the three levels of Spirit's Memory.

The Fiery World: Life begins in the Fiery World, the highest of the three. The Fiery World is the world where the Spirit's Thought is highly active, where it is represented by a constant motion of the finest vibrations of Fire and Light, creating the archetypes of life. To use a computer analogy we can see it as the 'operating system' of Spirit's Memory. In the Fiery World everything is together – past, present and future. No forms as we understand them, just the highest frequencies of the Fire of multi-dimensional Universal Consciousness – Infinity and Eternity instead of Space and Time. Also the first archetypal form of Matter of Spirit's Memory – the Prima Materia – exists there as a kind of 'primary plasma'.

Time, which is inseparable from Matter as the inner space of Matter's vibrations, begins in the Fiery World. Time begins as a reflection of the cosmic rhythm, 'the heartbeat of the universe' – a space helix, serpent, or wave of extreme density (call it 'zero' or 'infinity' – it is the same in the condition of Nothing/Everything).

Our highest, Causal body belongs to the Fiery World, and Spirit's Thought abides inside us. This is what is animating us living beings – the fiery all-dimensional Spirit Consciousness. This is the source of our own consciousness. Our three-dimensional brain is only a part of the mechanism of putting our fiery consciousness into action in the Solid World, where we struggle to apprehend any dimension above our habitual three.

The Fiery World is the world where we come into complete unity with all that is and into complete understanding of everything. "Of course, when entering the 'Vijl-Space' (Nirvana, or 'No-Time') it doesn't matter, with return or forever, a human being doesn't lose their individuality, Essence, energy envelope: so-to-say, fiery flesh." [17]

Once the 'Manvantara' (the active period of a manifested universe's life) comes to an end and 'Pralaya' (the period of rest) commences, the whole universe becomes united in the Fiery World, and all Matter collapses into a point of Nothing/Everything.

The Fine (Subtle) World: The Fine World is layered plasma, a kingdom of Light with vibrations of lower frequency than those of the Fiery World. Here the creative work of Spirit's Thought develops archetypes of concepts, laws, and scenarios of Creation. This is the world where liberated initiates, free from worldly karma and attachments to any material form, come to experience blessed unity.

Kalagia calls the Fine World – 'the World of Perfect Ideas'. This is a kind of permanent memory containing the archetypes utilized to construct the universe and all the mechanisms of its functioning. Light is preserving the Fire of Inspiration. Truth and Justice abide here. Fire and Light on this level of the Spirit's Memory are focused in the form of crystals of thought which have the possibility to manifest in the Solid World.

In our understanding, a crystal, an archetype of a form of life, is a closed structure form. Energy is moving within the crystal, following the 'lines of force' of Spirit's Thought which provides it with a balanced harmonious structure creating beauty. This beauty becomes evident when a crystal is manifest in the Solid World without distortion.

We do not perceive the Fine World with our physical body's sensors. We can reach some layers of its vibrations by the sensors of our finer bodies if we develop them enough. We sometimes appear in the hologram of the Fine (Subtle) World in our dreams, meditations, out of body travel, near death experience, and eventually after the death of our physical body.

There are lots of well documented examples of such appearances. Books are written about them, and many of us carry in our memory remembrance of such situations that happened to us personally or to someone we know.

So, how do the properties of the two finer worlds differ from those of the Solid World? People who have shared their memories of appearing in the Fine (Subtle) World usually notice that they didn't perceive time as they had perceived it before. This is because in the Fine World they could instantly change things by thought. For example, according to *The Holographic Universe* by Michael Talbot, they could immediately change their appearance – body and clothes – if they so wished. This is a world of thought forms where the creative activity of thought is evident. Everything is so fluid in the Subtle World that the wave of Time can be 'squeezed' to zero with no apparent effort.

When we discuss Time of the Fine (Subtle) World, we should have in mind that the linear sequential time that we are used to is an illusion of our perception. Time is non-linear; it is a wave with its flow and density. *Kalagia* speaks of the pressure of the wave of Time density that produces the energy field around and within the atom and makes life possible:

> *There are no leptons; there is Nag – the wave of Time density, and atoms and Worlds exist in the zone of its pressure. A wave after a wave follow, only Spirit knows the speeds of their movement; for Us they are moments of Our existence, as a standing wave of life is constantly changing because of the interchange of atoms scudding across the Infinity…*
>
> *The wave of Time density appears where Fire is released out of the energy of the Perfect Body of Spirit's condition. The pressure of the wave of Time density is acting in the Zone where multitudes of holograms appear and are scudding away from the place where they were born.*[18]

So, where Spirit's Fire of Consciousness is released, there the pressure of Time density brings the standing wave of Life to manifestation. In other words, the zone where Spirit focuses Its attention and releases Its Fire becomes a 'holographic focus' for the manifestation of Life.

Space also looks different in the holographic Subtle World. Anything can appear anywhere, showing the inherent quality of a hologram where the local possesses all the properties of the whole, and the number of dimensions rises towards the all-dimensional condition of the Fiery World. There are several layers of the Fine World, from the coarse vibrations of the Lower Astral plane, to the highest vibrations of the Higher Mental one reaching out to the Fiery World.

The effortlessness of actions – the result of 'squeezing' the wave of Time in the Fine World – justifies for us the necessity of Solid World conditions for our experiences. We need the illusion of linear sequential time in order to make efforts: sow causes, reap the effects, and learn our lessons. In this way we refine our consciousness and bring the precious gifts of our spiritual achievement to the Universal Consciousness of Spirit.

In the electro-magnetic, by its nature, Solid World, we need the 'resistivity' of the objects and experiences to make the current of electro-magnetic energy flow. In their absence the flow would simply 'short circuit' and collapse.

The Solid World: In the Solid World, Matter is visible to physical eyes, and Spirit and Its Fire of Consciousness are invisible, though maintaining their presence within the manifestations of life. It is the World of our everyday waking consciousness.

Space in the Solid World is seen as three-dimensional. This conforms to our vision of living in a world consisting of solid bodies, liquids and gases, which in their turn consist of particles such as molecules, atoms and subatomic particles. The descriptions, methods and formulas of classical Newtonian physics reflect such vision, which proves relevant and useful for dealing with our everyday life in the Solid World.

The Solid World is defined by *Kalagia* as 'a World of solidified concentrated Thought' and 'the foundation of Creation'. It is the arena where Perfect Ideas are being tried through choices and decisions and following lessons. It is the World where the Fire of Creation remains packed and hidden within the solid forms, as energy is inside an atom.

When this Fire is released, knowledge appears and becomes our human treasure. According to *Kalagia,* the human essence comes into the Solid World to gain the energies for its spiritual achievements.

In the Solid World we commonly perceive time as a straight

line from the past to the future. Because of this illusion, we become dependent on Time to fulfil our dearest desires and dreams. In a way, we become 'slaves of Time', of our own illusion of its linearity.

However, it appears that even in the Solid World we sometimes manage to feel the reality of the spiral of Time. Its spiral essence is manifest in everyone's experience of life, with its ups and downs, mistakes and lessons, returning to the same point, but on a different level of understanding. A moment comes when we become capable of 'squeezing' it to the point in Space and Time where everything comes together. This is an instant of passage through the boundaries of Worlds, and of power over Time, which would be one of the highest achievements for a human being on his/her spiritual Path.

Continuity and its Applications: Stemming from the Law of Unity[19] is the Principle of Continuity – everything is in Unity in the Fiery World of the Creator, and everything is a continuing flow throughout the three Worlds of existence.

Continuity means that the archetypes recorded in the permanent memory of the Fine World can be manifested as the images of the Solid World to the extent the Observers of this World are capable of envisaging such archetypes. In the following chapters we will discuss how these images appear in nature, music, and the human body.

Continuity means also that the Universal Order of Spirit's Thought translates into the mathematics that science applies to describe the continuing flow of energy throughout the Worlds, its transformations and processes that it instigates. The lower the World, the less dimensions of description.

For example, Albert Einstein in his famous formula: $E=mc^2$, actually discloses the relationship among the three parts of the mechanism configuring the universe: light, energy and mass (matter). We can also rewrite Einstein's formula as $m=E/c^2$, to demonstrate that in certain conditions energy turns into matter. Contrariwise, matter turns into energy, and the coefficient of the equation is velocity of light squared.

Also, Schrodinger's wave equation, a partial differential equation, describes how a physical system's wave function evolves over time. It was developed to determine how the quantum state of a material system changes with time. And it could be applied or adjusted for the application for the standing wave of Time density.

When we expand our vision, either to the macro-world of cosmic bodies or to the micro-world of subatomic particles and waves (where classical physics sometimes cannot even discriminate a particle from a wave), then it may be helpful to expand our conscious perception into the higher Worlds (out of the Three Worlds of Existence) and take this continuing flow with its attributes and mechanisms into consideration. Our united universal consciousness allows such effort to succeed.

We may find in this way a feasible explanation of such cosmic phenomena as 'Dark Energy' or 'Dark Matter', and a new outlook for a 'Black Hole'. We may review the definition of Matter as we will do later in the book, and also find answers to some our personal difficult questions.

The questions that arise in physics about the real nature of time and about the strangeness of the behaviour of the smallest of elementary particles, leptons, also often cannot be answered from the standpoint of the classical knowledge of the three-dimensional Solid World alone. Even the great physicist Albert Einstein was troubled with the 'spooky action' of observation that could "instantaneously influence what happened far away without involving any physical force" and could cause appearances of small objects in the experiments of quantum physics. [20]

Meanwhile, these phenomena receive their natural explanation when we are aware of the holographic nature of Matter and the 'nonlocality' as a property of a hologram. Holographic objects are actually not made out of solid matter. They rather represent foci of thought or energy fields resulting from the pressure of the wave of Time density.

Though the three Worlds co-exist in any point of Space and Time, they differ in frequency. That's why observation in different Worlds requires different sensors. We can suppose that when an elementary particle (as it sometimes happens in quantum physics experiments) disappears from the eyes of the observer in the Solid World, that maybe it changed its frequency and could be found in the Subtle World by an observer equipped with sensors suitable for that World.

In any case, our observation of the picture of the world and what we see and consequently manifest depends on how we understand the world – as a single world of solid objects or as the co-existence of the Three Worlds vibrating with different frequencies.

The Monad in the Three Worlds of Existence

Vibrations, Holographic Code and the Monad: Fire and Light express themselves as electromagnetic waves or vibrations of different frequencies. Starting from the highest frequency vibrations of the primeval Fire in the Fiery World, right through to the lower frequencies of Matter in the Fine and then Solid World.

Everywhere we look we can see the reflection of universal archetypes: a creating trinity and a multi-dimensional wave or a spiral form. The creative Trinity consisting of Spirit, Its Fire of Consciousness and Motion from them and in them was at the beginning of Spirit's creativity. Two whirlwinds of energy spiralling in opposite directions were 'weaving' the 'lines of force' for the archetypes. These were the mechanisms of the Fiery and Fine Worlds, and we also can see the same rules and patterns in the Solid World. For example, in the genetic code of protein production they are represented by the triplet codons and the double helix, the dominant structure of nucleic acids such as DNA and RNA.

Prima Materia, the Primary Matter, containing the information code of Spirit for the first, Fiery World, goes through all the three Worlds as the information-energy field, or holographic code of creation. This code is put into action by the energy of the primeval Fire ('Fohat'). The information-energy field of the Fiery World becomes the holographic code for the development of Secondary Matter with its Fine (Subtle) World of vibrations, lower than those of the Fiery World. The information-energy field of the Fine (Subtle) World defines the form of the Solid World bodies as their energy body.

Every form in the universe is defined by its holographic code. Spirit's Consciousness animates its forms and enables them to interact with each other through the process of differentiation of universal consciousness. Universal consciousness presents itself in different forms yet remains united in the Fire of Spirit's Consciousness.

The term 'Monad', stemming from the Greek 'monas' or 'oneness', is widely used in different schools of esoteric knowledge. Monad indicates the universal consciousness which is undergoing this process of 'unified differentiation', and also the centre of consciousness of every living essence, the result of differentiation.

A circle with a central dot, as in fig.3a, is the simplest symbolical image of an undifferentiated Monad, which expresses the undifferentiated wholeness or unity of the World of Spirit's Consciousness and the seed of Creation within Infinity.

H.P. Blavatsky in *The Secret Doctrine* emphasizes the spiritual nature of monadic differentiation. It is the Monad – the consciousness centre – that becomes the centre of consciousness for different forms of the material world – from the mineral kingdom through the vegetable and animal kingdoms, and right through to humans:

> *In short, as the spiritual Monad is One, Universal, Boundless and Impartite, whose rays, nevertheless, form what we, in our ignorance, call the 'Individual Monads' of men, so the Mineral Monad – being at the opposite point of the circle – is also One – and from it proceed the countless physical atoms, which Science is beginning to regard as individualized (The Secret Doctrine, v.1, p.178).*[21]

The Monad, the One, starts turning into individual centres of consciousness as soon as the essences (as observers in the holographic mechanism) start to see themselves as individuals with separate consciousness centres which can interact with other individual consciousnesses.

Blavatsky states:

> *The 'Monadic Essence' begins to imperceptibly differentiate towards individual consciousness in the Vegetable Kingdom. As the Monads are uncompounded things, as correctly defined by Leibnitz, it is the spiritual essence which vivifies them in their degrees of differentiation, which properly constitutes the Monad – not the atomic aggregation, which is only a vehicle and the substance through which thrill the lower and the higher degrees of intelligence.*[21]

Having in mind the process of holographic materialization, we can understand this conditional differentiation into interacting units of consciousness which we also call 'Monads' in the higher Kingdoms of life, such as Animal, and especially the Human Kingdom. The Monad is most pronounced as a human unit of consciousness, so the connotation of this term relates most often to the 'Human Monad'.

Kalagia calls the human Monad – the 'Reactor of Evolution' and says:

> *We can symbolically call the Monad an energy capsule filled with Information and Fire which make the human Essence. Inside the Monad – the Fiery Seed of the human Spirit is in a slumbering condition. We have to awaken it, more precisely – a human being has to be awakened to the activity of the Fiery Seed of their Spirit.*[22]

The Human Monad has a most developed archetype, in the image of the first Spirit's hologram, the Logos, which serves as a holographic code in all the Worlds of Creation.

Kalagia thus depicts this holographic code of the human Monad:

> *Atman-Koshcham or Monad, with its envelopes containing chakrams, fiery wheels, and the Silver Thread going through inside of them, is the holographic image in the Prima Materia of the One All-Uniting Preeternal Spirit. This image is the holographic code of the development of Worlds.*[23]

The human energy body contains all the features mentioned, such as chakrams and the main 'lifeline' of the Silver Thread. The human body as we know it in the Solid World follows the holographic code contained in its energy body, not only as a whole, but also in all of its parts. For example, the human ear resembles the human embryo which shows us that the whole human physical body and its part, the ear, developed following the same code.

Spirit remains simultaneously in all of Its creations as their Source, their fiery core. All living beings are manifesting Spirit's archetypes as observers, in their way, in accordance with the state of their consciousness.

We know that a human being can consciously observe its own potential hologram and the potential of the World produced by Spirit. A human being is constantly changing the manifest hologram and moulding it according to the personal and collective state of consciousness. That's why human manifestations differ from each other and differ also from the initial, perfect, God-like potential for a human being.

Reality and Illusion in Human Life: Human beings are carrying within themselves all the three worlds, though out of these worlds, we usually see only our solid material body. Nevertheless, though it is rarely acknowledged, this 'solid' body cannot exist without the other two, i.e. the fine and the fiery ones.

Life is not accidental; it is a process, where all the elements are interconnected and necessary in order to provide for the existence of the whole system. In this universal system, Spirit, Its Fire of Consciousness, and Its laws of life are real and unchangeable. However, the holograms of Matter are changeable and depend on information participating in the manifestation of the hologram. This information is constantly changing, and it makes Matter, in a sense, an Illusion.

The cycle of material life is explained in this excerpt from *Kalagia*: "Illusion is finding life in the Fire of Spirit as Its Memory, it becomes denser until it reaches its material state, becomes deposited and...destroyed, developing into the state of Spirit, i.e. turns again into the Fire of Spirit, which is Spirit." [24] And a human being plays an essential role in this grand play of Spirit.

The Rays of Light form the hologram of a human being (see fig.1). The first is the 'reference' Ray – the pure Ray of Spirit from the Fiery World within a human being. The second Ray – the 'working' Ray – is the Spirit's Ray, which goes through the Fine World in the human energy body as the 'Silver Thread'. It starts as Logos in the Fiery World, and continues in the energy body of the Fine World, absorbing the colours and shades of human experience throughout the lifetime of the Solid World human body.

The Spirit's 'reference' Ray, being simultaneously the 'reconstruction' Ray, initiates the potential making it perceptible for the Observer in the human being. As the Observer manifests the hologram in co-creation with Spirit, it becomes manifested in the Solid World as a material holographic projection.

When Spirit withdraws the 'silver thread' from the solid body, this body dies and is destroyed in one way or another. The Fiery and Fine World essence is free to develop further without its shackles. If there is a purpose, the energy body will build a new solid body for its reincarnation.

A solid body is necessary for human experiences, learning and development. Through this body's experiences we choose which colours and/or shades of our potential we are ready and willing to manifest. In this body we are separating order from chaos. In this body we can either become a toy for societal influences and manipulation or, by listening to the inner voice of conscience ('The Voice of the Silence' according to H.P.Blavatsky), we can become the master of our destiny. And we are able to create harmonious wholeness out of the experiences of our multiple lives, reaching the state of our archetype – the Logos. This is the spiritual Path towards enlightenment.

Creation of Creators

The 'Crystallized Light' of Prima Materia: The Fire of Spirit's Consciousness gives birth to Light. If we saw this with our physical eyes, we would perceive this primordial Light as complete darkness because it doesn't spread. It is contained inside Spirit's Memory as invisible energy or energy turned into primeval matter – Prima Materia (the 'Mulaprakriti' of Theosophy and 'crystallized light' according to H.Blavatsky).

Modern science states that the expansion of the universe is accelerating because of 'dark energy', which means for us that the process of the universe's creation (manifestation) is still going on. As defined in "The Urge to Live", 'dark energy', in our understanding, is Light that doesn't propagate until Time is released. Where Time appears, there we see the accelerated expanding universe, the ongoing process of creation.

As Light is forming holograms of Matter, it loses its intensity and the frequency of its vibrations and eventually 'freezes' in Matter. On its way Light is creating holographic worlds, living worlds, where the changes in motions of life are providing food for the Fire of Spirit's Consciousness.

As we depicted in our book *The Laws of Life,* Logos originates Monads – Spirit's units of Consciousness.[25] These are conditionally individualized units because in reality they are still united in Spirit's Consciousness. These units, Monads manifested by Logos, become the Observers continuing the process of manifestation. They, in turn, participate in the creation of the top of the 'Hierarchy of Light'[26] – the Hierarchs of Cosmic Bodies, Stellar and Planetary Phenomena.

When their manifestations become visible to us, we see their light and movements reflecting the cyclic – spiral archetype of the 'serpent' of Time.[27]

For us, in our solar system, the Sun is the primal source of light and life. Our Sun is the manifestation of the so called 'Sun behind the Sun' – the archetypal Sun produced by the Creator's Thought. [28]

The Perfect Pattern: The Creator, Spirit, is sending out rays of light – an infinite multitude of rays. A ray, when passing through a phenomenon (or experience), creates a hologram of Matter. Then an observer of this hologram manifests a new hologram by providing the observer's own experience.

All the components of the process of manifestation are expressed in vibrations, where the frequency of vibrations diminishes as they are forming material holograms. The Idea of highest frequency – the Thought-form of the Creator – starting in the Fiery World, receives its experience in all three Worlds: the Fiery, the Fine and the Solid World. Spirit's Thought creates Beauty. The universal archetype of the world, the Perfect Picture in the Memory of the Creator, in all its immense Beauty could be observed in information crystals of Prima Materia formed of the highest vibrations of the Fiery World (some spiritual teachings call it 'Akasha'). When eventually vibrations come to a certain low frequency, they 'solidify' and become perceivable for human sensors as Matter in the Solid World.

The Creator is Creating Co-creators: So, who are they, these observers and Spirit's co-creators? As we can see in the picture, fig.5, Logos becomes the first Spirit's co-creator in the progression of observers who co-manifest holograms of life in the three Worlds of Existence.

Holographic manifestation of Logos' Thought creates Observers, co-creators, in the Fine (Subtle) World. However, nothing can replace for Spirit the experiences of the Solid World because in the Solid World these experiences happen in Time and Space. So, Spirit, through and with the Logos and the observers of the Fine World co-create observers in the Solid World. We humans are such observers. As *Kalagia* puts it, "the human being on Earth is a concentrate of Lord's Will projected to the Earth by the ray of Logos through the Fine World".[29]

We are observers of everything created by the Creator's Thought, and in our way are co-creators of the holograms of Matter. Our experiences participate in the process of creation, and our collective observation is forming the picture of creation.

Who participates in the collective observation which is forming the picture of creation, besides us humans? Every living essence, every being endowed with a Monad, a unit of consciousness: A Star, a Planet, a plant, an animal, small or large. They are all endowed with consciousness as long as they are

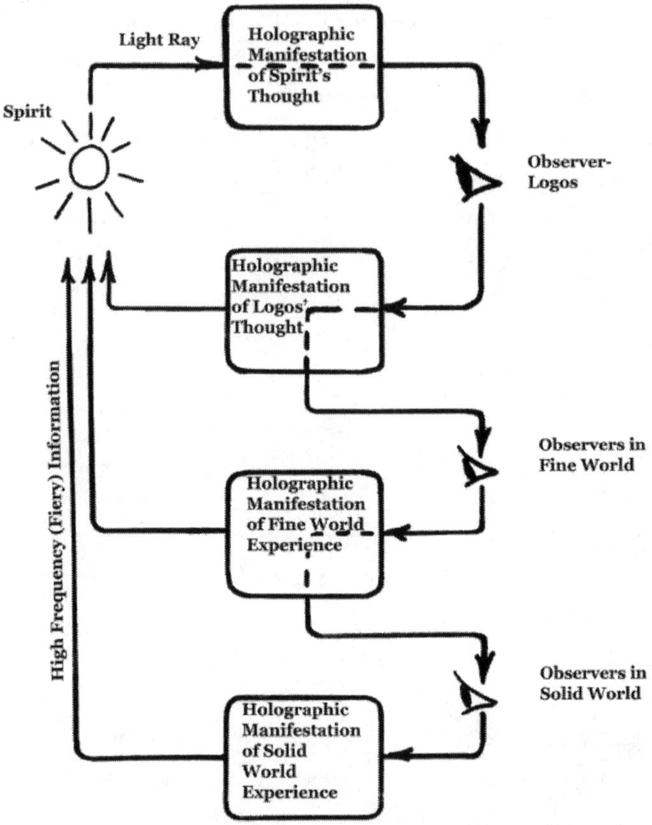

Fig.5. Creation of Co-creators

connected to the Fire and Light of Spirit's Consciousness. Also, because all these creations are manifested as holograms, their every part reflects the properties of the whole and may act as an observer. In chapter 3, in "Holography in Cells", we will discover an 'observer' in a ribosome, the decoding element of a cell.

As *Kalagia* states, we are created in a process of "Creation of a Creator by the Creator of Creation". It is the only possibility for the Creator to know Itself, to see Its reflection – through the process of Creation of a Creator. Every time we experience, observe and create, we are reflecting a facet of the Creator, and ourselves become created as creators. The Creator, by going through our experiences, knows Itself thanks to our creativity.

We humans are most eager to know the ways of developing our skills as Creator's co-creators. We are doing this through experience, learning and becoming aware of what we have learned. As we can see in "The Urge to Live", fig.1, this experience is forming the 'working ray' and is actually the source for any new holographic manifestation.

The whole process is going on and is recorded in the Memory of the Creator within which we all abide. And vice versa, every human being, Spirit's co-creator, is carrying the Creator within. Because of this we humans belong to all the three Worlds: the Fiery, the Fine (Subtle) and the Solid World, and inside us we have the potential of a Creator to manifest the beauty of the perfect picture of Creation. So, on the one hand we are created by Spirit and manifested through holography, but on the other hand we are also observers and co-creators of the hologram of life.

Our creations depend on the state of our consciousness. In the Fine (Subtle) World when our mind is in a pure and clear state, it creates crystals of thought. And in the Solid World we experience, create and recognize solid matter in the shape of atoms and their congregations, which in fact also consist of electromagnetic light energy vibrations.

For millennia, we developed our intellectual abilities through experience and experimentation in the conditions of the Solid World. We are used to seeing the objects of this world only and to believing that nothing else exists. Our intellect generalizes and creates abstract concepts for the explanation of our phenomenal world. But human consciousness didn't always work this way.

One of the oldest people on the Earth, the Australian Aborigines, may serve as an example of a different way of thinking and observing life. It seems that they are positioned for observation somewhere on the boundary between the Fine and the Solid Worlds. "Their Time is circular, not linear, as each generation relives the Dreaming activities".[30] Their 'Dreamtime', to which they constantly refer, shows all the attributes of the Fine World as of the world of 'dreaming' which 'clothes' itself in a body of the 'waking' world. According to Robert Lawlor's book *Voices of the First Day,* the Aborigines understand the creation as a 'movement of consciousness from dream to reality'.[31]

They have generalized images for the archetypal world of the 'Dreamtime', for example, the 'rainbow serpent' representing the shapes and energies from which they supposedly originate.[32] In contrast, they have multiple words for the same objects of their waking life which differ for different situations. For example, they can name differently "the same fish or animal at different stages of its breeding cycle".[33] In their way, the Aborigines showcase every living thing as an observer who manifests their waking world with its particularities conforming to the observer's individuality. Of course, there should be a separate word to name such individual creation!

According to Lawlor, the Aboriginal people also refer to their great ancestral beings as 'those who created forms out of their own vibratory energy'. We can understand these ancestral beings as the 'observers' of the Fine World, who actually co-created our Solid World out of the Fine (Subtle) World. As Blavatsky writes: "...the first human stock was projected by higher and semi-divine Beings out of their own essences".[34]

How We are Growing as Creators: Spirit is formless, but It creates the forms of matter. And It remains in every created form, in its 'seed', as its essence. *Agni Yoga* calls the Spirit in Its creation 'spirit', and says:

> *The rays of space penetrate into the seed and intensify the potentiality of the spirit; only by this impulse can the spirit create. When the potential of the spirit comes in touch with the ray of Cosmos, the spirit-creativeness is asserted. Thus, each form is*

> *generated by the spirit and the cosmic ray. How little people ponder over the grandeur of creativeness! How little understood is the thought which is tensed by the cosmic ray! How little understood is the Cosmic Might in all its manifestations! When humanity will understand that creativeness comprises cosmic seeds, it will assimilate the cosmic rays.*[35]

We may call the change of consciousness as a result of this impulse 'revelation'. By revelation we, all the animated forms, are growing and revealing our essence to the world. By revelation, a flower is opening its petals. We people by revelation become aware and able to create. The more we are open to the spirit within us the more often and eloquent are our revelations, the more we reveal the very essence of Spirit in and from ourselves.

Each of us individually and all of us collectively are manifesting the hologram of a living and active world. As creators of our world, we would like this world to be as beautiful and beneficial to us as its potential is in the perfect picture of the Creator's Thought. Somehow we are often faltering in our efforts to do this and produce something that we don't like at all, but why? Probably, because our attention is constantly fluctuating between the perfect picture, where we find our revelations, and the matters of the Solid World, where we are still going through our experiences and learning. The current state of our development becomes reflected in our creations.

The crystal of Universal Thought is based on all possible experiences in the world, contained as crystallized ideas in the perfect picture of the Creator's Thought. If we were able to manifest this crystal, we would manifest it immediately, by the power of our thought manipulating rays of light. But usually, while acting in the Solid World, we haven't experienced all the experiences upon which the perfect picture is based and therefore our manifestation is rarely complete.

Nevertheless, we often imagine ourselves as those who

can unilaterally decide how the picture should be manifested according to our desires. Very often we thirst for power because in our delusion we think that we only lack Power to become the Masters of the picture of the world. We desire to possess the abilities of true alchemists without actually becoming them. We then distort the picture of our own manifestation, produce karmic effects, suffer from them and start learning through suffering. Literally we learn by 'banging our heads against a brick wall' until we learn better.

Of the three Worlds to which we belong, our innermost, Fiery World, possesses and supplies us with the most powerful creative energies of the highest frequency. These energies can make miracles. While our attention is focused in the Solid World of Matter, our thought vibrations are not always reaching the high frequencies of the Fiery World, and such thoughts usually lack the energy of Fire for their fulfilment. That's why we cannot see the immediate results of our thoughts. We have to enforce them with effort, labour, and skills for a required span of time. We are going through our trials, making choices and exerting ourselves to turn the chaos of our lives into order. In this way, we connect with our inner Fire and receive an ability to produce creations of immortal beauty.

As we humans depend on time, our consciousness is limited in its ability to reflect the complete, perfect, timeless picture of the Creator. Nevertheless, we are always carrying this picture within, in our spirit. After incarnations of experience and exertion, while trying, suffering and learning, we are gradually perfecting our abilities. Eventually, our lifestream comes to those shores of Eternity where our reflection of the Creator is complete and perfect, equal to the universal archetype of the Logos. We have reached the state of a realized archetype. We then become the Creator.

As *Agni Yoga* told us:

> *In their earthly condition people cannot always notice the beneficial growth. All growth comes with pain, with lightning, and in storm, and only a perfect consciousness can perceive the radiance of Truth. Every evolving thinker must bring to realization his understanding of Truth while on Earth.*[36]

The Masters speaking to us through Agni Yoga are themselves examples of perfect consciousness. They assumed responsibility for assisting us earthly beings while we are growing spiritually and bringing our understanding of Truth to realization.

Any ability comes with assumed responsibility. An Ascended Master is already an archetype of human perfection and possesses the ability to create crystals of perfection in no time, out of Light, directly by thought. The closer our thoughts are to the frequency of the Masters' thoughts, the more we are receptive to their messages.

Though there is likely to be an eternity of future trials for us humans, even now we are able to see the crystals of the beautiful, perfect, timeless picture of the Creator's Thought in the 'ore' of suffering and effort. They usually come as destiny, as revelation, and as a 'gift from the heavens'. However, such moments wouldn't be possible at all without the thorough preparation at the times of our effort. So, no effort disappears without a trace, nothing is in vain in our life. That is how the co-creation of creators with the Creator is proceeding. This is living alchemy in action!

Chapter 2

Manifestation of Nature

How Nature is Created

What exactly is Nature?: It seems that all of us know what it is, but we wouldn't find the exact definition of 'nature' in most encyclopaedias. Usually we perceive it as everything that is around us and isn't produced by us. Wikipedia provides us with another clue: "The word *nature* is derived from the Latin word *natura*, or 'essential qualities, innate disposition', and in ancient times, literally meant 'birth'." Which points us toward the feminine aspect of the Creator, the creative force ('Shakti').

We can also probably relate to 'nature' the name 'Neter'. This is what the ancient Egyptians called their gods and goddesses, understood as principles or powers of Nature. Goddess Neith, 'a weaver of primordial light' [1], is one of such powers.

Whilst we are living through our experiences of daily life and subsequently learning and growing as creators, Universal Nature provides us with its own examples of Beauty. All our natural surroundings: rivers, trees, flowers, animals and birds can serve as such examples. If we suppose that the ideas of such Beauty were born in the Consciousness of the Creating Spirit, then how did these ideas come into manifestation?

Light creates the potential for the ideas to be developed into the perfect pattern, the archetypal matrix, which forms the body of the Fiery and Fine Worlds. The archetypes of this matrix can then be manifested in the Solid World in which all living beings consciously exist and act at the present time.

We humans, together with all other forms of life, as observers, manifest all that our individual and collective mind can perceive and accept at a certain point of our development. In this sense, we are co-creators with all living beings and with the Creator in all of them. Together we are 'drawing and painting' this miraculous 'colouring book' of Nature. Manifesting Nature in form, colour and sound, using the archetypes to form its manifestations.

The Archetypes are Shaping Nature: So, what is there in store in the archetypal matrix for the shaping of Nature? Let's take a look.

One of the first archetypes in the Spirit's Memory is the Spiral form. A spiral form, starting from the 'Serpent' of Time, reflects the cyclic nature of the universe's life, including the gradient of its development. "As above, so below". We find manifestations

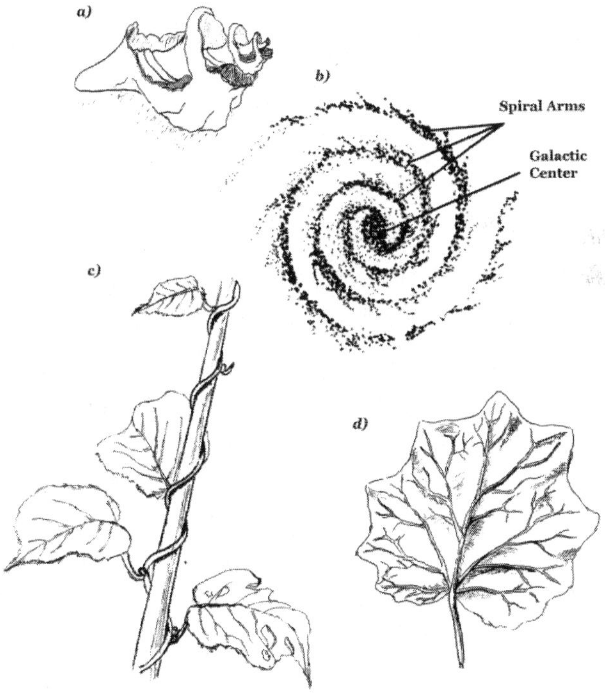

Fig. 6. Archetypes in Nature: a) A shell, b) A spiral galaxy, c) Coils of kiwi plant, d) Branching leaf veins

of spiral structures everywhere in Nature, from a shell on a beach (fig.6a) to a constellation of stars (fig.6b). We also find them in the spiral forms of embryos, horns, ears, fingerprints and many other forms of natural objects. We find a spiral in the shapes of tornados, the movement of planets, and in all types of trajectories of different objects, beginning from the wave of Light energy.

We find the spiral shape also in all kinds of plants. In the patterns of bracts in pinecones and small dry fruits in the kernels of maize. In the coils of climbing plants (fig.6c). These coils can be right-handed or left-handed, and however you try to change their direction, they will always eventually return into the position natural for the plant because they adhere to a certain archetype. We find the spiral archetype in the pattern of leaves on the stem of any plant as it provides the plant with the best conditions of exposure to rain drops and to the sunrays.

The archetypes reflected by Platonic solids are also widely applied in Nature's shapes. For example, crystals of salt are made up in the form of a cube. Capsids (protein shells of viruses) take the forms of different Platonic volumes; very often icosahedral forms (20 triangular faces, see fig.4 b).

The archetypal patterns of the whole repeat in its parts. We can see the pattern of branching in a tree in the pattern of veins on every one of its leaves (fig.6d). The pattern of branching itself is also an archetype. In some plants we can clearly see the pattern described by the mathematics of the 'Fibonacci sequence'.[2]

The Fibonacci sequence starts from two numbers: 0 (Nothing/ Everything) and 1(Unity, Monad). Their sum is the third to make up a Trinity: 0+1=1. If we continue adding every two last numbers together, we will have the series of the Fibonacci sequence: 0,1,1 ,2,3,5,8,13,21,34,55,89,144,233,377,610,...

Michael S.Schneider writes:

Look at these series. At the first glance we see a chain of numbers. But look beyond the visible numbers to the self-accumulating process by which they grow. The series grows by accruing terms that come from within itself, from its immediate past, taking nothing from outside the sequence for its growth. Each term may be traced back to its beginning as unity in the Monad, which itself arose from the incomprehensible mystery of zero.[3]

The repetition of the pattern, the "principle of ongoing growth from within", and "the pulsing rhythms of growth and dissolution" are common properties of growth in Nature.

The Fibonacci sequence leads us to the discovery of the proportions of the 'Golden Mean': the ratio which we get by dividing any number of the Fibonacci sequence by the previous number (for example, 3/2, 5/3, 8/5, etc.) [4]. The Golden Mean ratio is usually symbolized by the Greek letter Phi (Φ). The ideal Φ is equal to the endless number 1.61803398875… or rounded off to 1.618 or to 1.62. The further we go in the Fibonacci sequence, into the bigger numbers, the Φ is closer to the ideal.

If we divide a line at its Golden Mean, then the ratio of the whole line to the longer section is equal to the ratio of the longer section to the smaller section. If we bend a line into a half-rectangle, applying the Golden Mean, we get the basis for a perfectly balanced rectangular frame.

Nature applies these archetypal proportions in all of its shapes and structures. They provide these structures with perfect balance. The parts of the human body are balanced by the proportions of the 'Golden Mean'. The navel divides the body, the brow divides the face, the wrist divides the arm/hand, and the bones of our hand are divided at the Golden Mean ratio.

It appears that the 'body' of our Solar System (having in mind the distances of the planets from the Sun) is also structured roughly at a Golden Mean, if we take into account the asteroid belt between Mars and Jupiter. This is an approximation because the planets are constantly moving against each other and distances change correspondingly.

Where people in their creativity endeavours follow the archetypes of Nature, they also come to most harmonious results. For example, the ratio of the 'Golden Mean' has been applied by the great architects and artists of the past – from the Egyptian pyramids to the Gothic cathedrals and paintings of the Renaissance. The masterpieces based on this ratio are always powerful, harmonious, and therefore beautiful.

Archetypes in Music: Thomas Carlyle, a Scottish philosopher (born 1795) once said: "If you look deeply enough you will see music; the heart of nature being everywhere music". We can say also that at the beginning of the universe there was music because the universe began from rhythm, and rhythm creates music. Percussion music, for example, is based only on rhythm.

With the addition of tone, melody became possible, and the possibilities of its manifestation received additional dimensions which are governed by the same laws and mechanisms of Creation. The 'shape' of music, its order, is represented by a musical scale. The seven notes of the diatonic scale correspond to the seven layers of the human Monad: "The Human Spirit has its seven bodies – envelopes of the Monad, within which the Source of the Spiritual Life – the Spirit – comes into being".[5]

M.S.Schneider writes in his book about the sacred philosophy of the ancient people and their notion of music's order:

> *The seven note music scale is meant to model the hidden side of a macrocosmic design, the universe ruled by mathematical harmonies of music. The scale structure implies that the universe emerges from absolute divinity, descends through a seven-stage celestial hierarchy, and returns to absolute divinity.*[6]

Musical scale can be likened to a climbing plant – it 'climbs' from the lower tones to the higher ones in groups of seven plus one note, called 'octaves' – where the eighth note repeats the first one but an octave higher. Which makes the experience a complete revolution of a spiral, or rather, helix, to be more precise. The spiral archetype, with the number 'seven' at its core, when it turns into an octave, becomes 'eight' (a number from the Fibonacci sequence).

When we are enriching the diatonic scale by adding to it the half-tones of sharps and flats, as it is done in a piano keyboard, then we receive a 'chromatic scale', which is also carrying the signs of the Fibonacci sequence. There are five half-tones in this scale, and on the piano keyboard, the black keys are situated in groups of 2 and 3 keys – all these numbers are the numbers of the Fibonacci sequence.

Creating Vibrations – Colour-Tone Radiations: How does Nature receive and express all its variable qualities? How is the 'colouring book' of Nature actually 'coloured in'? The primordial Light is White. It contains the potential of the full spectrum of Light frequencies reflected in the basic seven colour spectrum of a rainbow and also additional countless tints and shades. In other words, all the colour (or tone) vibrations, or the colour-tone radiations, are in the colour white, ready to be perceived by our senses in conformity with their frequency. As the Australian Aborigines believe, the great 'Rainbow Serpent' has shaped the earth and everything on it.[7]

From the direct spiritual experiences of the anonymous Russian prisoners incarcerated underground and calling themselves 'X7', we know that Rays of Light create nature in conformity with the qualities of the Ray's colour and tone.[8] Why colour and tone? How are they connected?

Sound and Colour both are made of energy waves, vibrations that originate in the primeval Fire (Fohat). They reverberate in human Fiery, Fine and Solid bodies. Our solid bodies use sensors which are different from the sensors of our fine energy bodies, where colour and sound can be perceived simultaneously as 'colour-tone radiations'. In the Solid World some of the vibrations we perceive as light and some as sound.

We still perceive light through our eyes as an electromagnetic wave which travels with the speed of 299792458m/s. Sound, however, is perceived by our ears from a mechanical pressure (acoustic) wave. This wave affects our ear drums through the medium of air, so in the earthly atmospheric conditions our perception of sound changes.

An example of our different perception of simultaneously produced light and sound can be our response to a powerful electrical discharge in the atmosphere, simultaneously causing a lightning bolt and thunder. We can see the lightning bolt practically immediately but we only hear the thunder after some time because the pressure wave has to travel through the air and its velocity is lower than the velocity of Light.

There is no contradiction in comparing colour and tone. We can say that Light and Sound are colour-tone vibrations or waves in the common spectrum of frequencies. We humans only perceive them as separate phenomena because we perceive them through different sensory organs (eyes or ears) supplied with different means of perception.

The Bilble says: "In the beginning there was the Word". As we understand it, this Word, which in Theosophy we call 'Logos', is on the one hand the first Thought-form of the Creator – Spirit, and on the other hand – the Sound (soundless sound for us humans because our physical senses don't hear it). However, our energy bodies have the ability to perceive the vibrations that are inaccessible to the physical senses.

There are some people who can 'see' sound or 'hear' light with the help of their invisible sense organs which are part of our Fine World body. When we rise to a heightened level of spiritual development where we develop our currently latent sensory abilities, light and sound merge into colour-tone radiation as they did for the 'X7' in their underground cave: "All is light, light is radiation, radiation becomes colour and/or tone. The two are equivalent."[9]

The 'X7' mention Solar Substance as the source of all colour-tone radiations in our solar system. Through the Essence of the Sun, a Hierarch of the 'Hierarchy of Light', we receive the Rays and all the endless possibilities of the creative energy, which is the Source of Nature's manifestation.

Our physical eyes perceive sunshine – a part of the spectrum of colour-tone radiations, whereas the sensors of our Fiery, Fine and Solid Bodies combined can perceive a sum of the rays of the Essence of Sun, of the life giving Solar Being.

We see the blue of the sky. As the 'X7' state, 'blue' is the colour which contains a sustaining quality. Our planet Earth couldn't maintain its integrity if the skies had any other colour. In addition, all the other colours are equally necessary for the whole of life on our planet as it exists at the moment. (In reality, we perceive radiations not only through physical eyes, but through our skin and all visible and invisible organs and cells of our being.)

So, when we are looking at the shining blue of a summer sky, we are feeling good because we receive the sustaining energy of the colour 'blue'. With this colour we receive our strength – the quality of the colour 'blue'.

According to the 'X7', 'mother of pearl', with irradiations of rose-pink, is the colour of love, the foundation of life. Therefore this colour can best reach human consciousness. The quality of Peace has a perpetually harmonious, strong, multi-coloured beam.

The colour 'yellow', according to the 'X7', means freedom, joy, and relief from tension. We all feel the enlivening power of the colour 'green' and know the green colour of chlorophyll which enables plants to absorb energy from light through the process of photosynthesis.

Nature, its rocks, trees, flowers and any other living being reflect the initial thought forms in their shapes, colours, and other qualities. The 'X7' give us an example of a tree, which reflects the quality of 'tree-ness' – something like strength, endurance and stability.

Creation Begins from Responsibility: We humans are conscious observers and can affect the phenomenon we observe. In this way, we are co-creators with Nature.

Consciously or otherwise, everyone is creating and co-creating with Nature. Perhaps, when the majority of people can understand that they can create their world this way, they will stop fighting for a piece of the Earth. They will understand the Earth as a living, conscious, manifested hologram of reality. They will learn to respect and endorse others in their unique ability to create Beauty. And this will stop wars and conflicts that presently afflict the world.

Co-creation of Beauty begins from Responsibility. As *Agni Yoga* declares:

> *In rejecting responsibility, man deprives himself of a most wonderful feeling and of a personal, affirmed, creative force. Man, transmuting the fires of Space into forms, cannot cast aside his responsibility. Each form created in Cosmos will correspond to the form of the spirit. All generated forms will correspond to the forms created by the spirit in its ceaseless striving toward perfection. The feeling of responsibility must raise the cosmic creativeness.*
>
> *When the spirit develops in the direction of Infinity, then the feeling of responsibility acquires the power manifested by the Creators of Cosmos. Begin to realize personal responsibility and strive toward cosmic responsibility. There is the step of personal responsibility, the step of responsibility for human thought, the step of responsibility for human evolution, the step of responsibility for a better future. When thinking can be*

perfected, it will be possible to say that the time of construction of a better future is near.[10]

Everyone's consciousness is united with the consciousness of others. On the other hand, we are unique because each of us fulfils their role in creation, or co-creation with Nature. We are working with the colour-tone Rays that we are able to perceive in every lifetime. We are forming our manifestation of Nature by observation and actions. We are cooperating in the process of creating the world in which we live, peacefully or in conflict, depending on the state of our consciousness, with other living beings such as plants and animals, and other forms of life.

Gradually, with our developing spiritual qualities, we learn to widen our colour-tone abilities. Through successive lifetimes and initiations, through the transmutation of our bodies in all the three Worlds, we can ascend to the state of an Ascended Master, a Master of a Ray, and assist others following on behind.

What is Matter?

When talking about the creation of Nature, one usually has in mind the nature of the Solid World. For centuries, humanity was working at the task of creating and refining a reliable science that would be able to define and describe every phenomenon of this world.

How Science Defines Matter: We are used to thinking that we live in a 'material world', i.e. in a World of Matter. But what is Matter?

Some examples of scientific definitions of Matter are: "Matter is anything that has mass and takes up space" (in chemistry); and "Something that has mass and exists as a solid, liquid, gas, or plasma" (in physics).

These definitions are based on 'mass', but what exactly is

'mass'? If we try to answer this questions by using Einstein's famous formula $E=mc^2$, we get $m= E/c^2$, which means that mass can be derived from the ratio of energy to the speed of light squared. It appears that Einstein actually discovered that mass depends on the amount of energy and the speed of light. Consequently, the scientific definitions of matter could also be rewritten as: "Matter is anything of energy, depending on the speed of light, and taking up space", and "Matter is something of energy, depending on the speed of light and existing as a solid, liquid, gas, or plasma".

Would it be complete enough to define matter? Maybe it is possible to expand our options of defining matter by looking for a definition that would be true for its substance throughout all of the three Worlds of existence? We will try to answer the question "what is Matter?" on the basis of our understanding of its nature and the mechanism of its manifestation.

How We Understand Matter: In the cycle of Life, we regard Matter as ever existent. However, it becomes manifested only where Time appears.

Some explanations of the nature of Matter we find in *Synergetics, the Introduction to Kalagia,* by Alexander Naumkin: "The source of Fire is Plasma – Prima Materia", and "Light is born in Fire, Plasma carries Light out of the sphere of Fire's action. Plasma is Matter that is carrying Light".[11]

We understand the process of "carrying Light" as holographic manifestation of Thought through the medium of the electromagnetic vibrations of Light. The primary Plasma of Prima Materia, which is also Spirit's Body, Soul and Memory, is at the start of this process. It becomes the innermost substance of any Matter consisting of different frequency electromagnetic vibrations.

The process of manifestation of Matter begins in the Fiery World, from the Plasma of Prima Materia, and continues in all layers of the Fine (Subtle) World until Plasma comes to the condition of gases. In the Solid World, Matter manifests as plasma, which is ionized gas, gases, liquids and solid bodies. The energy of the electromagnetic vibrations of Light remain at the very core of the substance of Matter, in any of its particles; and Einstein's formula confirms it.

So, we can expand the physics' definition of Matter and define it as – *a holographic projection of Thought by Light manifested in Space and Time as different frequency electromagnetic vibrations of plasma, gas, liquids, or solids.*

Reflection of the Process of 'Weaving' in the Matter of the Solid World: Spirit's Thought is creating perfectly ordered archetypes of Matter through 'weaving' them by electromagnetic lines of force. The results of this process, projected as Matter, appear in Fine and Solid worlds.

We can 'catch' manifestations of this process of 'weaving' of archetypes in the processes of our well explored Solid World. For example, in different times scientists observed regular patterns in periodicity of such properties of elements of matter as atomic size, valences, melting points, reactivity, and, conversely, inertness. Therefore they created different versions of the Periodic Table of Elements, showing us that the patterns of the elements of Matter are not created chaotically, but in a mathematical periodic order.

Scientists also found that 'elementary particles' – electrons – possess an intrinsic angular momentum and a magnetic moment. This property of an electron could have occurred if the electrons were spinning balls of charge, and therefore it was called 'electron spin'. As we understand, this might be a reflection of the spinning of the poles of opposites under the forces of repulsion and attraction, the mechanism of 'weaving the web' of archetypes.

Every part and particle of holographically manifested Matter possesses the qualities of the whole hologram. That's why we can see similar archetypes manifested in the macro- and microworld. For example, the archetype of the double helix of creation and evolution is working for the whole creation and evolution of universes and for the genetic mechanism defining the potential development for any living being on Earth.

Illusion and Reality of Matter: All living beings create their versions of holographical manifestations of Matter of the Solid World and, according to *Kalagia,* present its contents to Spirit as Illusion. These versions are all illusory because they represent personal experience and understanding drawn from experience and limited by the scope of the experience. Every manifestation is presented from a certain angle of observation and point of view. Therefore it can be and is presented in countless versions. However, this Illusion is an essential necessity for Spirit because it allows Spirit to see and express Itself in all the facets of manifested life.

We humans, in the process of our evolution, become co-creators of Reality with Spirit. Experience and learning, sometimes gradually, and at other times instantly, turns our Illusion into a crystal of Thought which is of Reality. Manifestations of crystals of Thought in our Solid World are always harmonious and beautiful. We can see them in Nature and in the highest achievements of human thought in arts, sciences and technologies.

Whilst most of our thought considerations will in time disappear, some of them could be of use as experiences for our learning. Certain crystallized information of high frequency vibrations of our thought remains in the memory of Spirit as the Reality of Its archetypes.

Co-Creation with Nature

Searching for Harmony with Nature and its Laws: Coexistence and co-creation with Nature has been the human way of life for centuries. In different parts of the Earth we can find numerous examples of simple and intricate methods and tools applied to this purpose. One of the most eloquent examples are the 'living bridges' of a village community in Northern India.[12]

Consecutive generations of villagers are building 'living bridges' out of the roots of living trees. They are tending to the trees and stretching their roots. As the trees develop, the roots grow, and people tie them together in such a manner that the roots eventually make a bridge across the river. People are sacrificing their time and effort for the village, for the community, understanding this as part of their responsibility and purpose in the life of the community. Though those who start the bridges usually will not live to see them complete, such is their sense of purpose and their altruism that villagers work tirelessly to create new bridges and tend to the old ones.

Nature has its own laws[13] and where we are working in harmony with these laws our co-creation with Nature produces Beauty. We can express it in a universal formula:

HARMONY (LOVE + WISDOM + POWER) = BEAUTY

This means that when our power, love and wisdom are applied harmoniously, the result is Beauty. We are naturally entitled to the powers of creation in accordance with our level of spiritual development. If we are not asking for more than our fair share of powers, then we should be able to apply these powers, in harmony with our love and wisdom, to the benefit of both humans and nature. This can be expressed through the Law of Responsibility which states – *we are responsible for whom we love and for what we love.*[14] As *Agni Yoga* puts it, our 'personal, affirmed, creative force' comes when we take on responsibility for that which we create.

Everything is united in its Source, and another law, the Law of Purposefulness, tells us – *only that happens and continues which is purposeful from the standpoint of Spirit.*[15] We are free to exercise our free will, live our life, create anything, draw plans and realize them, make mistakes and learn from them. Whatever we do will exist for a time or even for eternity, if it is in harmony with the unified picture in Spirit's Memory.

That which we call 'Nature' shows us great examples of such harmony. It calls us to follow suit and teaches us to search for our own purpose, with respect to the purposes of other living beings, Nature, Earth, and the Universe itself. If we learn this lesson, then we co-create with Nature harmoniously. If we fail to do so, then some unexpected and unwanted changes can occur. Human contributions to climate change are an outstanding example. If this is the case, the only thing we can do is to learn and try better. This is how our wisdom is growing, and it can be seen as the real purpose of life for a human being.

We all, as integral parts of the Creation, originated from the One all-permeating Consciousness that we call 'Creator' and remain in the timeless Creator's Memory in the Unity of existence, which defines the harmony of our coexistence. In this Unity, all of Nature's treasures – sunlight and the water of the oceans, and all the beauty and gifts of the mountains and forests – are available for our use. This is Nature's sacrifice to us humans and all the other living creatures on the Earth. To co-create in harmony with Nature, we have to match it. According to the Law of Sacrifice – *the Creator sacrifices Itself to the Creation; the Creation through its sacrifice develops into the Creator.* "[16]

Love of Nature calls us to expand the sphere of our responsibility. Expanding the sphere of responsibility automatically expands our consciousness. By expanding our consciousness and understanding of the innate forces of existence we gain influence over the elemental forces of Nature. Like Aladdin with his lamp, we then can put the 'genie' to work in order to fulfil our wisely understood purpose.

In the Creator's Memory the Perfect Picture is already uniting the past, present, and future as a harmonious whole. But we are given Time as an opportunity for the realization of this Picture. Out of this understanding comes our perception of interconnectedness and purposefulness of everything, and our ability to live and co-crate in harmony, as well as to sacrifice some of our superficial and selfish desires to this harmony. We will see the possible consequences of our actions with greater clarity.

We will understand the great power of ideas. When we plant an acorn, we can expect an oak tree to grow, given enough time. When we plant an idea, we should understand that we will receive a 'tree' of consequences.

Human Dilemmas when Dealing with Nature: Dealing with Nature is not a straightforward process. The number of people in the world is rapidly growing and has already acquired colossal creative and destructive abilities. We humans exist in constant need of new material resources and sources of energy.

Nature is calling for our co-creation by providing us the means for life. The sun, the wind, the tide – we can use renewable energy sources without any harm to the environment, either now, or in the future. What is more, our planet has already accumulated and secured for us substantial amounts of mineral and organic resources as a result of the billions of years of its existence and of the life of its innumerable inhabitants over the course of its long history.

The question for us humans is: do we wisely apply these resources? For centuries we could afford to only think of how to reach and use them. But now our powers have expanded to the extent where we ourselves have become a formidable force in our quest for Nature's resources, which requires us to act with love and wisdom. Now we can afford to replace some material resources with our intellectual resources and think of how to protect Nature, while providing for ourselves.

We have to recognize that our consciousness is not accustomed to always taking into consideration the necessity of this balance. We desire more power at any cost, and thus relentlessly dig into the bosom of the Earth to find it. We have in mind some aspects of our purpose and completely forget about the other aspects. Especially we forget about the future we are preparing for Nature and for future generations, including ourselves as reincarnating beings.

After digging up fossil fuels and turning them into energy for centuries, we have to deal with the local problems of air, water, and soil contamination. We have also to increasingly contend with the global problem of carbon dioxide emissions and the resultant climate change. Though humanity has already experienced radioactive contamination from damaged nuclear reactors, most notably in Chernobyl (1986) and Fukushima (2011), we nevertheless seldom have in mind the possible future harm from nuclear waste. At least such considerations seem never to be our priority.

Do our actions of today have a reason compelling enough to justify our risky strategies of energy production? The reality is that the race for more and more profit has become the driving force of our civilization. It makes us a wasteful civilization. We waste our resources in endless wars and preparing weaponry for future wars. We daily dispose of a colossal amount of technology, food, and other consumer goods, often turning it into toxic waste.

We are Learning: On a bright side – we are learning. We are already recycling much of what would become landfill waste. We are also creating and mastering new principles and methods of design that completely change the ways we are manufacturing consumer products.

For example, only half a century ago manufacturing machinery was heavy, power consuming, and extremely wasteful in every way. In the 21st century, an industrial designer receives an idea of a future product, creates prototypes using computer technology, and then sends the results to a manufacturer. The manufacturer can use a 3D printer to produce a 'made to order' and 'just in time' product for the customer. This is much less wasteful and power consuming, and clean. The only waste is bits of information that can be easily deleted from the computer's memory. Instead of heavy machinery, pure intellectual power is being applied. Instead of mass production, we are moving to individual creation employing intellect and imagination.

As our overcrowded world is based on building permanent habitats, we often clash with Nature's essential and sometimes vital requirements. For example, bushfires in the natural forests of Australia enable many plants to flower and reproduce. However, periodical fires make the coexistence of nature and people problematic. On the one hand, fire is a vital part of Nature's existence. On the other, people often suffer from natural bushfires just because of living in proximity to the forest.

That's why fire containment and forest recovery have become a constant balancing act for Australian environmentalists, firefighters, and the National Park Rangers taking care of re-vegetation. Much of their attention and effort is devoted to the prediction and planning of simultaneous measures for preventing the destruction of forests with all of their natural and human inhabitants.

Though using methods and equipment produced by modern science and technology, Australian environmentalists are actually searching for harmony of human relationship with the living bush. Harmony which the Aborigines of Australia cultivated for millennia. Though living as nomads, they felt any place where they stayed as "inseparable from the original activities that gave it form."[17] They harmonized their own movements and actions with the 'music' of the land.

We can understand the Aborigines as observers who acutely sense the hologram of their land, its fluidity and connectedness to their own bodies. In their mind they transfer the hologram of the 'Dreamtime' into their 'waking time' and choose their experiences carefully so that the harmony will not be interrupted. We also have our chance to live in harmony with the land if we learn to connect our vast modern knowledge with the love and sensitivity to the land of the indigenous people.

We are learning! There are always forerunners among us who have reached the state of natural balance. But 'en masse' humanity is still progressing slowly through stages of searching for power, constantly being out of balance, appearing at a point of crisis, and then being forced by circumstances to build its relationship with Nature anew.

There were times when man used all his powers to win back from Nature his morsel of food for survival. By developing his intellect he eventually tamed some of Nature's forces and made them serve his power. But his consciousness is still in a 'fighting mode'. He is mostly thinking of how to take the power from Nature, this time for profit, not for outright survival. Paradoxically, as pointed out by famous environmentalist, David Suzuki, wealthy people in our community invest in the destruction of Nature, so that they can afford country retreats where they can get away from the pollution caused by such destruction! It is 'a road to nowhere' because if we aim for the profit alone it inevitably results in a destructive outcome for Nature. The only constructive way is to come to balance again through deepening our love for Nature and accepting the responsibility for its state.

Our human mind has already reached the level of development where it is able to find necessary solutions for balancing our interaction with Nature, and its development continues. At this stage of our civilization, we need to understand that Nature is not only alive but conscious. It is full of sparks of life, 'monads', in every aspect of its manifestation. They are conscious and ready for co-creation with us.

We have heard of the experiences of advanced seekers on their spiritual path, such as H.P. Blavatsky, Helena and Nicolas Roerich, and other spiritual teachers. What we read in *Agni Yoga* and the experiences of the authors of *The Mysterious Story of X7* shows us, too, that our human abilities are unlimited. In the future we may be able to turn our minds into instruments of manifestation, relying only on our inner abilities and power, and Nature's freely available gifts. This would be an intense yet joyous state of labour, aimed at creating Beauty in whatever manifestation we choose, and simultaneously, a magnificent freedom of the flight of our soul.

Chapter 3

Manifestation of the Human Body

Manifestation on the Boundary of Worlds

We are United in Spirit's Body, the Soul: As *Kalagia* states: "Spirit's Body, the Soul, is simultaneously the Form of the Worlds (all of them) and the Form of a World (every one of them), which is the information-energy field of Spirit, i.e. Its Fire is Its Soul and the Soul of everything existing in It".[1]

Everything that we see in our Solid World starts from Spirit and has its origin in the Fiery World. Every form of Matter is made out of energy. Its contents, however it is seen and named, are made out of Fire, of the information-energy field of Spirit. This field is the indivisible all-permeating Body of Spirit. It becomes only conditionally divided into standing waves of vibrations of different frequency.

The standing waves of vibrations become holographically manifested and take on forms in conformity with the archetypes produced by Spirit's Thought. But these standing waves are of the same information-energy field that composes the Form of the Worlds. So, in reality, everything is united in the Soul which is the Spirit's Body.

The Matter of every World is made holographically out of waves of Light which comprise the electromagnetic information-energy field ('Spirit's Body') in motion. That's why *Kalagia* says that "the holographic sediment of Spirit's Information in Matter is Spirit's Body".[2]

Our Abilities of Manifesting Matter: How do we usually produce something material? We take resources from Nature (ore, oil, water, timber, etc.) and use it as the basis for manufacturing needed products. We use tools that are produced in the same way. Nowadays we apply sophisticated technologies, and sciences help us to improve and enhance these technologies and to invent even newer ones. All our projects require time to complete and sometimes this amounts to a lot of time.

We are used to thinking that this is the only possible way of creating anything material. That's why we have taken from Sir Isaac Newton's heritage only the part of science that describes such processes and turned away from the spiritual contents of the rest of it. And that's why we don't think any other way of developing our world is possible.

In this book we are talking about the natural processes of holographic manifestation of Matter out of the rays of Light applied by Spirit the Creator. Can we humans mimic such processes by our will and learn also to manifest material phenomena out of Light? It appears that we can, under certain conditions. H.Blavatsky, the principal founder of the Theosophical Society, could manifest bouquets of beautiful flowers just out of thin air.[3] The prisoners in the Siberian caves in the 1950s, calling themselves 'X7', learned to manifest things, including their own bodies (and 'unmanifest' – dissolve them) by will.[4]

We are natural Observers co-creating with Spirit (and Logos) in the process of manifestation of holograms. Most often we are doing this unwittingly and don't have control over the results of our manifestation. However, we can learn to manifest consciously by will. We can let the Ray of Spirit enter into our thought, turning this Ray into a 'working ray' of a holographic process and our thought into a manifested image (see.fig.1). However, here we may face the problem of Time. Time is the inner envelope of Matter and we have to be able to handle Time to maintain our newly produced manifestation for long enough.

This seems simple in theory, but in practice we have to initially develop ourselves to become capable of creating anything by Light. In the book *Saint Germain on Alchemy*, the famous mysterious alchemist writes about the alchemy of 'self' as a precondition for any experiments of manifestation.[5]

To receive the flow of Light through our thought, we have to ask for the ray of Logos. To handle Time, we have to get Power over Time. In *The Laws of Life*, we were writing about the long way of purification before any powers of manifestation can be received.[6] *Kalagia* says that the Soul should be kept clean from the Illusion of the Material World. And what is Illusion in that instance? *Kalagia* defines Illusion as "Suggestion of the Worlds of their own Contents to Spirit".[7]

For us humans 'illusion' would be a suggestion of our habitual conjectures as ultimate truth. This illusion can be our only perception of the possible truth of life until life itself provides us with an experience which completely alters our understanding from illusion to awareness.

The 'X7', in their underground cave, were devoid of any means of providing for themselves in conventional ways. They could only be silent, think and pray. Then they came into communication with a great Presence and received the Light that enabled them to experiment with the spiritual nature of Matter:

> *We thus became aware of our own Indwelling Presence, of the unlimited potential within, and the possibilities of creating a totally new world of form. This was a slow process, one in which we could not measure time. But before we could commit our minds to the truth that a man is a creator of his own world, we had first to eliminate from consciousness all preconceived ideas – all that held us to the pattern of the man of the world, all that bound us to the negative; for only the positive would respond to tonal qualities.*[8]

Creating with the rays of Light is our inherent ability. Even now, at the current stage of our development, we are able to accomplish it in some inspired moments. Though habitually we still need instruments and technologies of our Solid World, nevertheless our creative ideas and designs become manifestations of Light. Selfless ideas may produce a 'point of power' attracting Light for their manifestation. People naturally feel the attraction of places where such ideas are manifested for the benefit of everyone.

***Manifestation of Life in Different Worlds*:** According to *Kalagia*, "The Three Forms of Matter are three Levels of Spirit's Memory". Let's have a look at these Three Forms and at the ways in which Spirit's information-energy field transforms itself into the contents of the Forms of Matter of the Three Worlds of Existence.

What would we observe from the lofty heights of Spirit? According to *Kalagia*, we would observe:

> *Tertiary Matter and in it the physical planetary World and its inhabitants; Secondary Matter and in it – Worlds fluidic, ethereal, the inhabitants of which dwell between the planetary bodies; Primary Matter and the Fiery World in it, the inhabitants of which dwell on the plasma bodies of stars and in the free Motion of Spirit, Fire, Light and Heat.*
>
> *Each of these three forms of Matter's existence consists of seven parts. That is, the Solid World formed on the planetary bodies consists of seven parts: every planet has its seven bodies, from the solid one to the most subtle one; The Fine (Subtle) World, fluidic, ethereal, interplanetary, has its division into Spheres (space structures) where its Essences live; the Fiery World has also its fiery seven parts.*[9]

What are these seven bodies of each World and how can we understand them? We can see them as electromagnetic vibrations of different frequencies. These frequencies descend gradually from the highest ones of the Fiery World, lowering throughout the Fine and Solid Worlds.

As *Kalagia* says, "a targeted transfer from one World into

another is being realized through the substance of Time."[10] *Kalagia* also explains that the information-energy field of Spirit is actually the vibration, rhythm and intensity of the motion of the wave of Time density. So, the wave of Time density is carrying information. The flow of Time, which is the inner space of Matter, instils this information into Matter.

To manifest the holograms of Matter of every part of every World, Spirit needs Light and information for the 'working ray'. So the wave of Time density causes motion in the holographic information code of the forms of Matter and this motion is combined with the motion of Fire and Light in the code and with the motion of the code itself.

The movement of the Wave of Time density can be likened to a stairway with standing waves as its steps. It symbolizes the gradual way in which Matter moves away from Spirit, in groups of seven, i.e. by octaves. According to Michael S. Schneider, in ancient days "the earthly music was seen as a mirror image of the heavenly ideal descending from above".[11] In conformity with such understanding, a musical scale on the lyre was played starting from the highest note, the shortest string at the bottom of the lyre.

However, in the Timelessness of Spirit's Consciousness living Matter never parted from Spirit. On the one hand, Matter remains in Spirit's Memory. On the other hand, Spirit always remains within Its every form. From this point of view, the seven bodies of a unit of Spirit's Consciousness – the Monad – are just "seven envelopes – areas – organs of transmutation of Fire and Spirit into material, psychic and spiritual energies".[12]

So, if we look from the standpoint of Spirit, we can see a human being as 'a manifested Monad'. We see that we consist of seven bodies – envelopes – layers of vibrations belonging to all the three Worlds: "Human Spirit has its seven bodies – envelopes of the Monad. Inside these bodies, is the Source of Spiritual Life which is Spirit".[13]

The Display of the Wave of Time Density in Different Worlds: The Wave of Time density is moving from its centre outwards. The ways of its realization change on the boundaries of Worlds. In the Fiery World Spirit's Thought is 'drawing' archetypes immediately, in 'no Time', by means of its fiery lines of force.

In the Fine (Subtle) World, in the 'realm of all-penetrating Light', the archetypes exist as perfect ideas. The Fire and the Light of Spirit's Consciousness are gathered into focuses of crystals representing these ideas.

In the recent book by cardiologist Pim van Lommel, *Consciousness beyond Life,* a patient who had a Near Death Experience (NDE) during brain surgery, describes her experience of appearance and observation in the Fine World:

> *The closer I got to the light, I began to discern different figures, different people... The light was incredibly bright, like sitting in the middle of a lightbulb. I noticed that as I began to discern different figures in the light – and they were all covered with light, they were light, and had light permeating all around them – they began to form shapes I could recognize and understand. I saw many, many people I knew and many, many I didn't know, but I knew that I was somehow and in some way connected to them.*[14]

As we can see, the speed at which matter manifests in the Fine World is different from that of the Solid World. In comparison to our Solid World, in the Fine World the observer observes and participates in manifestation of aspects of life directly and almost instantaneously. That's why, when we appear in the fine planes of existence, our consciousness can travel through time and visit the past, the present, and sometimes even the future situations of the Solid World. It is possible to see the future as it appears in the holograms of the Fine World prior to being manifested in the Solid World.

***Manifestation of Human Monads in the Solid World*:** Procreation of humans is many times experienced and well researched. However, as *Kalagia* says, "There is something, secret and subtle, hidden behind the outward appearance".[15] This "secret and subtle" inner world – the cause and the spiritual contents of human appearance and disappearance – is multidimensional and complex. So we will follow *Kalagia's* lead when trying to explain the hidden causes and ways of human manifestation in the Solid World:

> *An earthly man is a manifestation (a concentrated scanning Ray) of the Man of Fiery Infinity. An earthly man is the 'Search' of a Cosmic Man. A Cosmic Man is 'Logos' – the Lord's Word, a Sphere of information – energy, stretched in the Universe as a common rule and as a special case – locally, above the planetary bodies. An earthly man is a concentrate of Lord's Will projected to the Earth by the Ray of Logos through the Fine (Subtle) World, where the Idea of Search is enclosed in ectoplasm (from Greek 'ektos' meaning 'outside' –transl. R&M) and becomes the Fine Body, which further becomes incarnated on Earth into the solid tissues of Nature.*[16]

Let's now investigate the process of incarnation – manifestation 'in flesh', on the boundary of the Fine and Solid Worlds – following *Kalagia's* detailed explanations. According to *Kalagia*, the Logos distributes Itself to the countless zones and areas of Creation, which in their turn are divided into units. These information-energy units have multitudes of holographic focuses – archetypes of Monads. The Monads are distributed in conformity with their functions.

Each Monad is unique, although there are Monads with similar functions. In the Fine World each Monad is developing its qualities which will be passed on to the incarnated human being. These perfect qualities will enable a manifested human being to fulfil the Idea of the Design for the Monad.

A Monad which is ready to be incarnated possesses properties of both sexes. The higher consciousness of Logos defines the future sex of the incarnated being, securing balanced homeostatic existence of the Fine World.

According to *Kalagia*, the Rays of three Essences participate in the process of conception: two already manifested Essences (Father and Mother), and the yet to be manifested Essence of the future child.

Conception of a new human being is possible only if both partners are incarnations of Monads belonging to the same information-energy unit of the Logos. Somewhere in Spirit's Memory, the laws of genetics are recorded, the karmic connections taken into account, and the future parents and children on the higher levels of their consciousnesses 'make arrangements' for cooperation in the Solid World. Then conception/manifestation begins.

As *Kalagia* declares, nothing can happen without Fire being released into the Space of Thought. "Biological flesh is a hologram (holography – animated hologram). Genetic code is also, according to its nature, the holographic outline of Fire on the Space of Thought".[17]

***Adam and Eve. Proliferation of Genetic Information*:** 'Ad' in Aramaic means 'One'. Adam Kadmon in the Qabalah (Zohar) means "One (Son) of the Divine Father, or the Creator". H.P.Blavatsky calls Adam "The first emanation of the Father-Mother of divine nature".[18]

In our understanding, Adam is the first human archetype created by the Creator in the Fiery World. Adam exists there in the sphere of the highest vibrations, where there is no Time. Time appears as the means of creating Matter, as its inner space (see "The Urge to Live"). Matter is developing following the coils of the spiral of Time (the 'Serpent' of Time). In this way, Adam is 'lured' by the 'Serpent of Time' into manifestation and further development in the Solid World, becoming a mortal, a 'slave of Time'.

There were several manifestations of Adam's archetype (called 'Manu' in theosophy), progenitors of human races. According to H.P.Blavatsky, the first three human races were androgynous, i.e. Adam's manifestation had the male-female properties. In the late third root race the separation of sexes began and in the fourth race Adam's and Eve's manifestations finally separated as sexes.

The properties and corresponding functions of the male and female aspects differ and complement each other. A female aspect keeps the 'database of information', which the male aspect replenishes with new ideas for development. The female aspect is also a representative of the primordial creative energy ('Shakti') necessary for realising the process of reproduction.

So, when a successive 'Manu' or 'Adam' fertilizes a female of an old race, she receives all the genetic information necessary for their descendants to start a new race. We will further explore how the genetic information works in the process of holographic manifestation of cells producing biological flesh.

Question: *"Do you mean that your description of conception is at work even in cases where a child is conceived as a result of abuse or in a successful case of 'in vitro fertilization' (IVF)?"*
Answer: *"The answer is – yes. If a child has been successfully conceived and born, it means that this event has its place in the Memory of the Creator."*

Holography in Cells

Manifestation of Organic and Inorganic Life: There is a difference in the ways in which the Mineral Kingdom and all other kingdoms of Life manifest the potential of Spirit's archetypes in the Solid World. In the Mineral Kingdom nature follows the structures of primary crystals of spiritual archetypes directly, which produces that which we call 'inorganic' minerals.

Not so with living beings of Vegetable, Animal, and Human Kingdoms – the so-called 'organic' life. In the Solid World, these active Kingdoms of life are manifested by more sophisticated archetypes. Actually, a hierarchy of archetypes spreading from the highest vibrations of Fire and Light of the Fiery World through to the lower frequencies of Light in Fine World, right down to the Solid World.

The basic structure of manifestation of living organisms in the Solid World is the Cell. The potential inherent in its cells should provide the organism with the means of manifestation, development, multiplication, self-restoration, and self-destruction, if necessary. The potential inherent in cells should also provide the diversity of living organisms with the means for interconnection and interaction.

This potential could be created by a mechanism which, on the one hand, encodes all the archetypal properties for a group of organisms, and on the other hand, provides for their individualization. Science has discovered that the potential encoded in a cell's gene structure serves all these purposes.

Let's have a look at the genetic structure of a cell with its particularities and working mechanisms and try to discover how these mechanisms reflect and employ the inherent archetypal properties which make manifestations of life in the Solid World possible.

The Cell's Holographic Mechanism: Dr. Peter Gariaev, Ph.D., member of the Russian Academy of Natural Sciences, member of the Russian Academy of Medical/Technical Sciences, member of the New York Academy of Sciences, and a pioneer of 'Wave Genetics', explains how genetic apparatus operates from the point of view of his research.[19, 20, 21] He describes the genetic apparatus represented by 46 chromosomes as a 'library' consisting of 46 'volumes' or 'books'.

Each chromosome, as a 'book', contains a text of instructions of how to build an organism. This text is made up of sentences (DNA) consisting of words (genes). The 'genetic alphabet' has only four 'letters' which appear in different combinations. The molecules of DNA are formed as double helixes, consisting of segments – genes.

So, how does the genetic apparatus work?

The organism first translates the texts, written in the 'DNA language', into the 'RNA language' (where DNA is deoxyribonucleic acid, and RNA is ribonucleic acid – a messenger that carries information needed for protein synthesis from the DNA to the ribosome). Then the texts become translated into the 'Protein language'. And our bodies are mostly made of proteins (not counting water).

After questioning some principles of the genetics basic model suggested in 1968 by Nirenberg and Crick, Peter Gariaev started his research of the principles and mechanisms of all of these genetic 'translations' applying the methods of mathematical linguistics and wave physics to the genetic apparatus. This research led him to the establishment of a new branch of genetics which is now known as 'Wave Genetics' or 'Linguistic Wave Genetics'.

A team of physicists and biologists led by Dr. Gariaev made some groundbreaking discoveries proving the holographic nature of the genetic apparatus. His findings provide us with some clues to the mechanisms of holographic manifestation of archetypes in Solid World through the inner workings of the genetic mechanism of living beings.

One of Dr. Gariaev's experiments proved that the genetic apparatus possesses holographic memory. His group reproduced the experiment of 'phantom leaf effect' performed by V. Adamenko in 1975.[22] The experiment demonstrated that if we place a small part of a freshly cut leaf in a high voltage, high frequency electric field, then it starts radiating, and we can see an image (hologram) of the whole undamaged leaf. The experimental image lived for 10–15sec and could be recorded on film.

Through further experimentation, Dr. Gariaev found out that molecules of the DNA of chromosomes, which are in fact liquid crystals, appear to be systems accumulating photons and radiating coherent waves of light. This means that chromosomes are actually working as lasers with a changeable wave length, which confirms the holographic nature of the chromosome's genetic mechanism. Dr. Gariaev speaks of wave lengths from 250n/m – ultra-violet light up to 800n/m – infra-red light. We can see that this is also the interval in which our body's sensors perceive light.

So, the genetic code system which is of a holographic nature should provide information for the manifestation of living organisms. Evidently, to build a viable organism, a plan (design) of its spatial structure is needed. Also, an embryo should be able to grow and develop into a mature entity, so the coordinate of time must be included in the code system.

The shapes of the organisms that would be manifested in Solid World, are given by the archetypes existing in the Fiery and Fine Worlds, which contain all the information of such a design (the potential, the 'perfect design'). Obviously, the genetic information encoded in the genome (organism's complete set of DNA), reflects the archetype, and when decoded, should be capable of manifesting the unique shape and properties of the organism in space and time.

Decoding the Genetic Code: Amino-acids are the 'building blocks' from which the proteins of living organisms are built. Their combination defines what kind of an organism will appear.

The decoding element of the cell is called 'ribosome'. Ribosomes are responsible for the correct choice of amino acids for the realization of the genetic description of the genome. Dr. Gariaev asks – how are they able to do this correctly even in cases when the information they receive is ambiguous?

In the language of linguistics, which science applies for the description of genetic information, this is called 'homonymy' – meaning a case of each of two or more words having the same spelling or pronunciation but different meanings and origins, for example, 'well' as 'good' and 'well' from which water is taken. How is it possible to recognize the correct meaning? People usually do it by consciously placing the word into context – the part into the whole.

The ribosome's judgment should be accurate and the choice every time correct from the standpoint of the archetypal plan of the space-time structure. Dr. Gariaev comes to the only possible conclusion about the conscious character of the genetic apparatus. In his understanding the ribosome is able to put the part into the whole, understand the whole context, and make a conscious choice when identifying an amino-acid. To make a correct choice, a ribosome has to understand the context of the potential of the hologram of an organism that it manifests, i.e. its archetype. From our point of view, this suggestion also shows the place of the ribosome in the cell's holographic mechanism as the cell's 'observer'.

Because of its holographic nature, a cell contains the same information as the whole organism. The complexity of the task of manifesting an organism requires every cell to process huge amounts of information. For example, a human embryo resembles many other living organisms. To make a human body out of an embryo, the specific proteins should be made exactly at the right time, and without mistaking human hands for an ape's hand, and so on. So the code system must be reliable. This obviously requires encoding a large number of aspects.

The human genome is made up of over three billion 'genetic letters' of information where the same 'letters' are repeated many times. Science currently knows the application as coding for protein production for only around two per cent of this information, and sometimes even calls the rest of it 'junk information'. Unlike humans, bacteria have only two per cent of genetic information not coded for creating proteins. Does it mean that bacteria are more effective in using their genetic information than humans?

Highly doubtful, as nature wastes nothing and no living organism can afford to contain anything 'excessive' or 'junk'. Every bit of information participates in the manifestation of life, which altogether makes the mechanism of manifestation so dynamic yet stable.

Gariaev's genome theory focuses on the remaining 98 per cent of genetic information as being the key 'intellectual' content of all cells of the human body, including the brain. As we understand, the 'intellectual' content provides the consciousness of an organism more sophisticated than bacterium (such as human) with the potential to develop its intellectual abilities, including speech.

Scientists are gradually discovering the mathematics of cell mechanisms. For example, a team of MIT, Harvard, and University of Massachusetts Medical School researchers "established that the genome adopts an unusual organization known in mathematics as 'fractal'. This architecture, called a 'fractal globule', enables the cell to pack DNA incredibly tightly, while avoiding knots and tangles that might interfere with the cell's ability to read its own genome. Moreover, the DNA can easily unfold and refold during gene activation, gene repression and cell replication".[23]

Fractal geometry deals with different forms which replicate themselves on various scales, where the same shapes repeat many times changing their scale and position of manifestation. A visual example of fractal shapes from the fine arts can be seen in Escher's famously enigmatic pictures. In cells, the repetition of 'letters' of genetic information allows them to participate in fractal algorithms similar to linguistic ones.

Fractal geometry could also be an illustration of the process of transferring information which was initially written in the highest vibrations by the Creator's Thought. The process uses the archetypal information and fractal templates which provide for the manifestation of living organisms in all their diversity through the workings of rays of light.

Through holography the images stored in cells, coded in genes, and then decoded by ribosomes, become manifested into what we perceive as the physical bodies of actual organisms with a potential to grow and develop. It starts from consciousness in the Fiery World and eventually manifests the organism with its physical body and its functions in the Solid World.

The Ways of Development: An archetype, especially a human archetype, expressed in a genome, besides its role in manifestation, contains a vast potential for development. In any genome, there are active genes, and there are latent genes. Gariaev says that new structural data reveals that the human genome is organized into two separate compartments. In his opinion, the active genes are kept accessible while sequestering unused DNA in a denser storage compartment and each chromosome alternates between regions of active, gene-rich DNA and inactive, genepoor stretches.

If activated, the formerly inactive genes will allow the organism to manifest new properties. In *The Secret Doctrine,* H.P.Blavatsky gives us an example of such changes by describing the development of animal and human eyes throughout millions of years of life on Earth:

> *In the beginning, every class and family of living species was hermaphrodite and objectively one-eyed. In the animal, whose form was as ethereal (astrally) as that of man, before the bodies of*

both began to evolve their coats of skin, viz., to evolve from within without the thick coating of physical substance or matter with its internal physiological mechanism – the third eye was primarily, as in man, the only seeing organ.[24]

H.P.Blavatsky refers to palaeontology's discoveries concerning the animals of the early Cenozoic age (65 million years ago) "whose fossil skull exhibits a perforation otherwise inexplicable". She also writes about "the chameleon, certain reptiles and even fishes" having the remnants of the 'third eye', now atrophied, but active in ancient times.

According to H.P.Blavatsky, humans of the first three races also possessed a well-developed and active 'third eye'. It is the eye connected to the pineal gland, [25] now retreated into the skull, which allows vision in the higher vibrations of the Fine World. In the process of Involution (materialization), when humans became more material, and less spiritual, humans needed to have spatial orientation in the three-dimensional material world for survival rather than expansive 'spiritual' vision of higher vibrations. People then became two-eyed at the expense of their third eye.

However, some spiritually evolved people have developed the abilities of their 'third eye'. H.P.Blavatsky predicted that as the result of the evolution of humankind, with its rising spirituality, these abilities will return to all of us. As observers, we will see which abilities we really need, which genes should be activated to manifest those qualities of consciousness necessary for the future development of Humanity.

The Universe in a Human Body

Differentiation of Universal Consciousness and Appearance of Matter: *Kalagia* tells us: "A human being is a model of the universe, and equally the Universe is a model of a human being;

both of them are illusions of the past in Spirit's Memory that form the united structure of the Cosmos".[27]

To discover how the universe translates into a human being's mind and manifest body, we will look at the universe through its attributes: rhythm as the ubiquitous quality, Universal Consciousness as the Source, motion as the precondition of life, thought creating archetypes, and light which creates life through holographic manifestation. We can see these attributes in any manifestation of universal consciousness and in its parts.

Universal consciousness is One and yet it differentiates into units – Monads. So a human being begins from the unit of universal consciousness, a Monad. As outlined earlier, a Monad has its seven 'bodies' or 'vehicles' of universal life energy of various vibrations.

The Monad's bodies can be understood as seven manifested levels of consciousness, or as 'layers' of vibrations. Starting from the highest frequencies of the Fiery World and lowering their frequencies while descending towards manifestation as a solid body.

These levels of consciousness form the seven energy bodies of a manifested Monad. They are held one within the other like a 'Russian Nesting Doll' (fig.7). The innermost level consists of the highest frequency of vibrations – Spirit, within every one of us. It unfolds from within followed by other bodies, including the solid physical body which would be the most outward one. There is a dimension of a complete human body where we can find another 'Nesting Doll' – a reflection of the first one. In the reflected 'doll', the solid body is surrounded by the layers of other energy bodies from the lowest frequencies to the highest ones.

The Monad's bodies can also be understood as manifestations of Matter on different levels of Consciousness if we consider our definition of Matter – where Matter is *a holographic projection of Thought by Light manifested in Space and Time as different frequency electromagnetic vibrations of plasma, gas, liquids, or solids.*

Motion acts within the creative Trinity of Spirit, Fire of Consciousness, and Motion from them and in them at the very start of Creation (see "The Urge to Live"). Motion turns the Fire of Consciousness into the flow of primeval energy and provides energy to every level of life. Motion is 'a must' in any living system. The very holographic process of creating living systems and organisms is based on the motion of the interfering light waves.

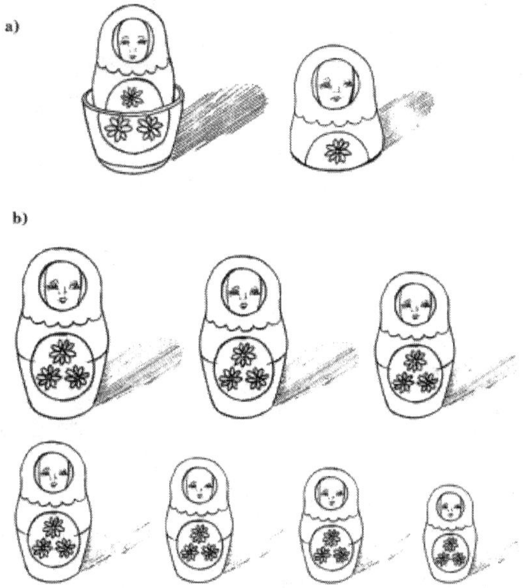

Fig.7. Russian Nesting Dolls
a) An Assembled Set of Seven Russian Nesting Dolls in the Process of Being Disassembled
b) A Disassembled Set of Seven Russian Nesting Dolls

Manifestation of the levels, or 'bodies', of the human Monad starts from the first level, which is the immortal consciousness of Universal Spirit. Its first form, the human Soul, the second level of the Monad is reflective of the Universal Logos.

The fiery Monad's high frequency energy vibrations gradually lower to the third level, or 'body', of mental activity. From the Fiery World, mental activity takes the ideas of the archetypes and manifests them into crystals of thought in the Fine (Subtle) World of the human mind. The actual separation of an individual also becomes realized on this level of mental activity as the concept of the 'higher ego'.

The individual 'higher ego' can choose what kind of thoughts it uses for its meditations, which eventually form human life and solid body. When human thought embraces the idea of Unity and the most unselfish and compassionate considerations, it may reach out to the level of awareness where its highest vibrations replenish the immortal contents of the Soul in the Fiery World. Where human thought is centred upon the gratification of animal instincts, human desires, and everyday life considerations, it lowers energy vibrations to the fourth level – 'body' of the Monad, on which emotions, wishes and desires can be expressed (the 'lower ego').

The fifth Monad's body, the so-called 'astral body', manifests Fine World forms of life. Human consciousness appears in the 'astral body' in a multitude of ways, including dreams, visions, and post-mortem existence.

The sixth body of the human Monad is the energy body closest to the physical body. It promotes the flow of 'prana'–energy necessary for life. Whether we are aware of it or not we constantly 'breathe in' this energy. The seventh energy level is manifested as the actual physical, solid body.

Out of these seven bodies, the highest three belong to the Fiery World. They are primary and immortal and remain unaffected by the processes of birth and death. Other bodies, with the exception of the last, solid body, belong to different levels of vibrations of the Fine (Subtle) World. The last, mortal, physical body belongs to the Solid World.

***The 'Silver Thread' and the Heart*:** All energy levels, or 'bodies' of the Monad are connected by the animating Ray of Light. Some spiritual teachings call this Ray the 'Silver Thread', or in Sanskrit 'sutratma' – the 'thread self'. All the holographic manifestations of a human body take energy and information from this Ray.

The Ray of Light produces an energy flow that moves along the spine. Spiritual teachings call the energy flow along the spine in the human body 'sushumna' (Sanskrit). The rotating energy centres, 'chakras' (wheels) are situated along the 'sushumna'. The spinal cord (a part of the central nervous system) and the spinal column surrounding it commence their existence from this energy and are feeding upon it.

In the middle of the spine, in the spiritual heart, the Ray releases spiritual Fire that ignites life in the body. Fire is being released within the 'heart chamber' belonging to the energy body. The energy of Fire initiates the major function of the physical heart in the 'universe' of a living being – to beat in harmony with the rhythm of the Universe. The heart starts beating and sending energy with the flow of blood into every organ and tissue of the physical body.

Modern medicine tells us that a baby's heart starts beating 18 days after conception. We can count the baby's physical life from that moment when the Fire in the baby's 'heart chamber' is lit, and his or her universe is connected with the Heart of the Universe, beating in synchrony with its rhythm.

People usually associate the human heart with love. The primordial energy of Fire in the 'heart chamber' is the energy of Love emanating from the very core of the Universe. If we don't put limitations on the flow of this energy and we are willing to share it with others, we are spreading Universal Love in the world.

Actually, every human being is carrying the energy of Love within. Life is impossible without the energy of Love which is primary to any other energy imaginable. Most people at the current stage of spiritual evolution of humanity cannot express all the aspects and colours of this Divine Love. We need to accumulate sufficient life experience, which we receive in a long string of incarnations, to become 'whole' and express this sense of wholeness in our Solid World.

Even the worst of people are craving and striving for love. Tyrants, criminals and abusers falsely assume that which they crave can be received from the outside world only. This is the reason why they try to extract this precious energy by any means. They do not yet realise their own ability to love and do not see the object of their possible love as deserving of their love. Sometimes their ability to love is restricted by earlier, especially childhood, events in their lives.

Throughout many lifetimes of experience, suffering and learning, people are gradually mastering their abilities to love. They start seeing and manifesting their love because they feel the stream of love energy which is flowing through them.

Consciousness and the Brain

Consciousness in Humans: Where does our thinking originate? Do we think with our brains alone or does our thinking originate somewhere else? This question has always fascinated philosophers and scientists throughout the ages. To mention just a few, in the 17th century, philosopher and mathematician Renee Descartes stated the opinion that there is a fundamental difference between the material substance of our body, including the brain, and conscious processes, such as thinking.

Max Planck(1858 – 1947), the famous theoretical physicist and Nobel Laureate, once said: "I regard consciousness as fundamental. I regard matter as derivative from consciousness. We cannot get behind consciousness. Everything that we talk about, everything that we regard as existing, postulates consciousness".[28]

As we understand it, workings of human consciousness can be found not only in the solid tissues of the brain but in the vibrations of all the seven levels – seven energy bodies – which make up the energy structure of the human body.

The solid body is the densest of the energy bodies. Modern science generally accepts that all the workings of consciousness are the product of the solid body's brain. At the same time, the research into Near Death Experiences by cardiologist Pim van Lommel,[29] for example, clearly shows that humans can sometimes think even when their physical brain is switched off. So what is providing the work of consciousness and the movement of thought in such a situation? The answer is – the other six levels, or energy bodies.

On which level do human beings think? This depends on an individual's spiritual maturity and orientation of the moment, and ultimately on the level at which a person can and chooses to 'fix their inner eye' as an Observer. The higher the level of vibrations the more powerful the thought.

There are special yogic practices and meditations which teach students to lift their minds to the higher levels of consciousness. In this way they can direct the energy of their energy body to rise from the lower 'chakras' to the higher ones. Creative aspiration or some life experiences also lead the Observer in us to intensely concentrate on the subject of interest, which in turn brings our thought to the higher levels of consciousness where we receive new understanding, ideas, or images. After meditating on the result of this concentration, the Observer can then put the newly gained understanding into action in the Solid World. New action means new experiences which lead to new understanding. By continuously repeating this process of 'concentration – meditation/understanding – action', the Observer turns it into a pattern of creative motion of consciousness. What role does the brain play in the motion of consciousness?

Holography in the Brain: As science has discovered, information is being transferred and transmitted in the brain by electrical signals called 'action potentials' along axons inside the neuron and through electrical and chemical neurotransmission between the neurons. This means that electromagnetic vibrations are directly involved in the process of thinking, even on the level of the physical brain. And the story doesn't stop here.

Neuroscientist Karl Pribram (1919 – 2015) came to understand the holographic nature of the brain's workings and formulated his findings in his pioneering 'holonomic brain theory'. In his research, Pribram collaborated for many years with mathematicians and physicists, one of whom was the inventor of technical holography, Dennis Gabor. The subject of Dr. Pribram's research and consequent theory was not only the brain itself but the whole central nervous system with all its sensors: auditory, visual, tactile and others.

In his essay, *What is Mind that the Brain May Order It?* (1997) Pribram described his discovery of a holographic mechanism for the manifestation of images by the central nervous system.

Pribram gives the example of visual sensors. When light penetrates to the retina of the eye, the information that it carries becomes translated into electrical impulses and is carried by the optic nerve to the brain. In the brain, the optic nerve transmits signals to the lateral geniculate nucleus (LGN) where visual information is relayed to the visual cortex of the brain. Supposedly, interference patterns of the holographic process caused by vision signals become constructed in both the nervous tissue of the retina and the cortex.

Pribram compares the interference patterns created by the rays of light with those made by waves that appear on the water's surface when a pebble is thrown into the water. He supposes that such distributed interference patterns "account for our remembering discrete items after brain damage. Sensory input must, in some form, probably as changes in the conformation of biomolecules at membrane surfaces, become encoded into distributed memory traces... The biomolecules would serve as neural 'holoscape' in the same way as oxidised silver grains serve the photographic hologram".[30]

Any incoming signals in the brain come as waves of certain amplitudes and frequencies. When they interfere, the interference pattern is a 'frozen record' of the nodes of interference among wave fronts. It contains the attributes (amplitudes and frequencies of these waves) in the form of a so-called 'frequency spectral domain'.

Pribram's theory states that "sensory systems perform a series of operations that combine to yield the Fourier transformation of the input signal impinging on the retina."[31] This means that the natural mathematics of the sensory systems conform to the procedure which French mathematician Joseph Fourier (1768 – 1830) described in 1822. A mathematical procedure called 'Fourier transformation' enables decomposition of a signal incoming as a wave into the frequencies that make it. The resultant 'spectral domain' of the interference pattern contains the information of instantaneous amplitudes against frequencies.

Pribram then enlarges the scope of his theory to cover not only sight but also other senses, including auditory and somatic (skin, muscle and visceral) sensations which are also processed as a spectral domain by the brain.

Pribram sees the function of the brain as a medium between consciousness and the matter represented by the central nervous system: "the brain to be the medium for the transformations that allow us to partake both of the 'ideal' and 'material'".[32] In these processes, all the ways of communication are working through the transformation of different kinds of vibrations (electromagnetic waves of light, acoustic, electrical and electro-chemical signals) akin to the rays of light which produce all the holograms of Matter.

Consciousness and the Brain: So, where is matter, and where is mind in this process? The answer is: matter as the result of holography is everywhere where there is a form. Matter is in any hologram starting from the first hologram of the Logos in the Fiery World. However, matter cannot continuously exist and live without the Spirit within, without Its Rays of Light and Consciousness – the primary source of thoughts.

The Rays of Light originate in the Fiery World and produce forms of matter through the universal mechanism of holography in Fiery, Fine and Solid Worlds. Through holography any thought-forms (forms produced by thought), including the archetypes, can be communicated from higher to lower energy levels. Along the way they participate in producing a multitude of holograms.

The Fire of thoughts is working in and with all levels of the human body, and produces new holograms by using thought-forms. Brain and the central nervous system as a whole is the main physical organ supporting this process. The resultant holograms are observed by a conscious observer, and turned into the observer's thought – experience, or thought-form, in turn providing the substance for a new hologram.

The thought-forms that can be found in the Fine (Subtle) World are the archetypal crystals. The archetypes, when manifested, create a most harmonious, perfect manifestation. We can see aspects of such perfection in the harmony of the human body, profound laws of mathematics, and the most enduring examples of music, art, and architecture. The archetypes also provide broad possibilities of development as in the universe as in the human body – "As Above, So Below".

The brain's functions serve the development, survival and functioning of the physical body: maintaining its homeostasis, processing sensory information, and making motor and other bodily functions possible. The brain also plays a crucial role in turning the images of thoughts into language constructions.

Holographic processes, as Karl Pribram discovered, are going on in the energy waves of the neural substrate, the neuronal connections, in the brain. The abilities of the brain are also provided and supported by the holographic mechanisms in cells. As we understand, these mechanisms use the 'excessive contents' of the genetic code. Though modern science still sometimes tends to call this content 'junk information', actually in the cell nothing is 'junk' and nothing is wasted. The pioneer in 'wave genetics', Dr.Gariaev, defined these 'excessive components' as the key 'intellectual potential' of human organism (see "Holography in Cells").

All processes that are going on in the brain can be understood as electrical or 'fiery' by nature, as even the brain's chemical processes are triggered by electrical signals. In all these processes, the solid structures of the brain, however subtle, play a subordinate role. Somewhat like a candle to the fire that feeds on the candle's substance.

All thought processes are going on in the energy flows and the brain tissue serves as an environment supporting these mental processes. When mental processes change, the tissue follows, hence the brain's 'neuroplasticity' meaning its ability to make changes in neural pathways and communications due to our experiences and changes in our environment, emotions and thinking.

Question: *"Can Artificial Intelligence (AI) eventually take over all the functions of a human being and even surpass humans in thinking?"*
Answer: *"Thanks to the great speeds and huge memory capacities of AI they may surpass and even replace human beings in everything connected with information processing. But the always new domain of creativity, either in sciences or in the arts, will still belong to humans because of their inherent inner connection with the Creator. On the way to creating the perfect design AI will never*

have the full potential and possibilities of human creativity and will not be able to equal, let alone surpass it.

Depending on their goals, human creators can create advanced mechanical aids or sophisticated weapons, which in some aspects will have greater abilities than their creators themselves. However, overall they will remain human creations of limited possibilities, limited by the same limitations that their human creators had at that time."

The Workings of the Brain and Neuroplasticity

Examples of neuroplasticity in science and medical practice demonstrate the electromagnetic and holographic nature of mental processes and show that thoughts are able to change the brain. Neuroscientist and neuroplasticity researcher, Norman Doidge, collected some of the most significant and compelling cases and described them in his books.

Neuroplasticity in the Books by Norman Doidge: In the book, *The Brain That Changes Itself* (2007), *Dr.* Doidge describes the experience of an inventive neuroplastician, Dr. V.S. Ramachandran, who was curing phantom pain using a mirror box to alter the patient's perception of their body's image. "Your own body is a phantom", states Dr. Ramachandran, "one that your brain has constructed purely for convenience".[33]

In his second book, *The Brain's Way of Healing (2015),* Norman Doidge cites a Russian born Israeli, physicist and healer, Moshe Feldenkrais: "The mind programs the functioning of the brain".[34] Feldenkrais came to the conclusion that from birth onwards our mind is continuously 'wiring' and 'rewiring' the connections within our brains, so that the brain is developing and changing its abilities. In his method of healing, "Awareness through Movement", he connects differentiated bodily movement with awareness. The method is taken from and very much reminiscent of the martial arts. As Doidge writes, "even thinking of making a movement triggers the movement, if even very subtly". As we understand, the movement of consciousness triggers the movement of the brain and body.

Development of the Brain: The connection between consciousness, brain, and movement, is a very important condition of development from the very beginning of human embodiment. The highest principles (bodies) of consciousness in a Monad are immortal, so a baby comes into the world with these principles already intact in the energy body. On the other hand, the newly born physical brain lacks the experience and training necessary to observe and manifest adequately. So the baby's mind starts training the brain to adjust to the new environment of each incarnation. Training is actually going on through experience connected with the fundamental property of the manifest universe – Motion.

In human life, we also note that Motion enlivens the brain and keeps it younger and healthier. For example, learning new skills keeps the brain active for longer in elderly people. Quite the opposite, stagnation without physical and mental challenge makes the brain become older and inflexible.

Norman Doidge states in his books that the brain is working by a rule: 'use it or lose it'. When we stop using a body part, the brain's part affiliated with it doesn't receive normal stimuli from the disabled body part and either shrinks or is reallocated to serve other body parts.

The main property of a hologram is that any part of a hologram repeats the whole. With brain, three interconnected holographic processes together support the ability of brain cells to fulfil the functions of other cells from other sections of the brain where these sections are damaged, removed, or not developed.

First, the whole brain and every cell is manifested holographically. So the hologram of the cell contains the same information as that of the whole brain. Second, holography is applied by the brain as a whole for processing the incoming information and developing the hologram of life as brain perceives it. Third, holography works in each and every one of the nervous cells.

In his book, *The Brain's Way of Healing*, N. Doidge corroborates Pribram's hypotheses and conclusions. He writes that "thoughts, memories, perceptions and skills" are encoded in "cumulative electrical wave patterns".[35] He even compares these patterns with "the musical score" which brain needs to "play its music".

The brain, then, can be seen as a channel or a mobile and malleable tool for the transmission of electromagnetic waves. Consciousness controls this channel, applying its instrument – thought. Some thoughts manage the processes inside the brain: some connect it with the all-encompassing electromagnetic process existing in the Creator's Memory and in every living thing.

New Ways of Healing: One of the major points of Doidge's books is that the central nervous system, including the brain, is sensitive to energy-based stimulation. Such neuro-stimulation can be provided by "light, sound, electricity, vibration, movement and thought".[36]

Out of all these stimulations, thought energy is the finest, and therefore the most effective one. This understanding can have very far reaching consequences for humanity's progress, as spiritual as scientific. It also has profound implications for medicine and healing processes of all kinds.

For example, Doidge describes the experience of psychiatrist turned pain specialist, Dr. Michael Moskowitz, who found a way of getting rid of chronic pain through visualization. Whatever part of the body we feel is in pain, the signal of such pain is produced by the nervous cells of the brain. When a person visualizes a painless brain or manages to turn their attention away from pain towards some other more positive sensations, pain is reduced, and in some cases, where willpower is strong, may eventually disappear altogether.

N. Doidge writes: "It reminds us that the sense of the body in pain is dynamic – it is being remade all the time, depending on visual input. It shows that altering the visual imagery of the body can modify pain circuits".[37] Perhaps it also shows us the great potential that we have as human observers to change the condition of our physical bodies by thinking.

The basic vibrations of light, sound and electricity are the major components of our energy bodies. That's why they are so efficient when being consciously applied. They have a history of application for healing going back thousands of years to the healing departments of ancient Indian and Egyptian temples and to the healing practices of indigenous peoples around the world.

Many treatments of modern medicine are already based on vibrations of a different kind. We can expect that the therapies of the future will inevitably adopt and widely use the clean and natural ways of healing and self-healing by thought, light, and sound vibrations. Physicians will hopefully become more like teachers and instructors, teaching their patients how to apply these vibrations in their home environment, and how to adjust this environment for a healthy life.

The mechanism of such healing is based on the cooperation of Thought, Energy and Creative Action. The origin of this Trinity is the initial creative Trinity of Spirit the Creator, Love Energy as the means of creation, and Creativity as the process. Healing starts from the intention of thought and uses energy from different sources. The rest depends on the creative application.

Chapter 4

Development of Life in the Universe and in Humans

The Cycle of Creation and Evolution

Creation and Evolution: Since Darwin in the 19th century, creationists and evolutionists have argued about how life started and developed – by 'creation' or 'evolution'? We suggest – by both, creation and evolution. The process of evolution is hand in hand with the ongoing process of creation. Both of them – creation and evolution – are indivisible parts of the complex rhythmical process of the development of consciousness and manifested life.

If we reduce the united process to 'creation' alone, then the whole mechanism of evolution is eliminated. Conversely, if we support the opinion adopted by modern science, that it is all down to Darwinian 'evolution', we come to life seen as a linear progression from the past to the future. This leaves us with an unknown beginning, ending, and without any particular reason for its existence. We may ask questions: "How does life begin so that it then 'evolves'? To which condition does it eventually 'evolve'? What are the inmost reasons for life to develop the way it does?" However, the answers we find through research of only material aspects of our world still give us a fractured model, if at all, because we are missing the vital connections between the different constituents of existence.

In comparison, a cyclic process containing both 'creation' and 'evolution' provides us with a wholesome model. To suggest such a model, we have to understand and redefine its constituents – the concepts of 'creation' (turning it into the concept of holographic manifestation) and 'evolution' (expanding the commonly accepted concept, meaning only the physical development of species, and turning it into the unfolding of consciousness resulting in physical development).

In this model 'creation' is the phase of Manifestation of Spirit's Thought where Spirit is 'in-wrapping' Itself into the forms of Matter. We call this process 'involution'. We understand 'evolution ' as the development of Matter into Spirit through 'un-wrapping' the Spirit from within its forms. The unfoldment of what is Within. We understand it as development and spiritual maturation of the 'co-creators' unfolding Spirit from the fetters of Matter and developing their conscious understanding of unity with Spirit and in Spirit. So, real 'evolution' is transformation of consciousness through the work of thought resulting from experience.

Within the great cycle of the Manvantara (Manifestation) and Pralaya (Rest-Time) of the universe, multiple cycles of creation and evolution of 'standing waves' of life are ongoing: cycles within cycles, within cycles. Within the Universal Manvantara, there are lesser manvantaras, such as that of a solar system, planetary manvantaras, and even lesser ones, such as of man.[1]

Every holographic process of creation starts from the archetype which can be produced only by a Consciousness capable of producing archetypes. This archetype serves as a 'blueprint' for the genetic foundation of the manifesting species of living beings. Then co-creation begins (see "Creation of Creators") and the experiences of consciousnesses of the manifested observers lead them to change their holograms, adjusting them to the conditions of the Solid World environment. The manifested holograms can deviate from the archetype as far as it is possible, creating the opportunity for the appearance of a new archetype.

Meanwhile, the process of 'evolution' is already under way. From the very beginning of 'involution' – 'in-wrapping' into solid forms – we are evolving, developing on the way and 'un-wrapping' the inner spirit from within the manifested forms.

While we are going through our revelations, we are producing crystals of thought and high frequency (fiery) vibrations which become part of Spirit's Consciousness (see "Creation of Creators, How We are Growing as Creators", and fig.5).

We understand Darwinian type 'evolution' as a reflection of the development of species through the development of their consciousness and adjustment to the conditions of the environment. This type of development usually takes place within the species. It can affect sizes, colours and some other characteristics of a given species. However, the manifestation of any completely new species requires a new archetype, and consequently a new cycle of creation and evolution. That's why involution and evolution are constantly going on in a cyclic-spiral motion.

Perhaps, the first archetypes at the start of the universal 'Manvantara' were not specialized enough to suit all manifestations of Life as we see it today. In the continuing process of creation-manifestation, more specialized archetypes appeared, separating different groups of living beings in conformity with their life purposes.[2] In the process of their evolution, changes of consciousness led to subsequent changes in physical bodies. Any development of forms comes out of the development of consciousness, which is co-creating with the Consciousness of the creating Spirit.

In the process of 'involution', the all-encompassing Spirit is creating holograms of Matter and animating their forms so that It becomes involved in the life of these holograms. These animated forms of the Solid World are not aware of their origin in Spirit, and of their unlimited possibilities. They have to 'evolve' to the heights of understanding and of completely expressing their Monad, the unit of universal consciousness. They have to return this way to Spirit, to eventually become all-encompassing in Spirit. As we understand, this would be their 'evolution'.

So, the complete and ongoing process of Creation happens

through processes of 'involution' and 'evolution' of Spirit. 'Involution' we understand as the process of materialization of Spirit's Thought, Its differentiation and creation of Matter. Spirit's differentiation creates living beings that carry Spirit inside of them, folded in the envelopes of their energy-matter bodies. These living beings would become Spirit's co-creators ("Creation of a Creator by the Creator of Creation" as *Kalagia* says) and would make evolution possible.

As for Spirit, as for every Monad as a unit of consciousness, this process begins from the condition of Nothing/Everything and after the Creation (Manifestation) transfers into the phase of Evolution, development. Evolution goes on until all the manifested and animated forms of Matter, through the process of involution and evolution, come to the state of Spirit, to the condition of Nothing/Everything. 'Involution' and 'evolution' together make a cycle of development, so the next cycle can start on a resultant higher level of development.

The processes of involution and evolution are going on with every one of us. We are feeding on the energy – information of the higher Worlds producing holograms of Matter – this is our involution. We are co-creating with Spirit, learning and spiritually developing on the way – this is our evolution. It is beautiful when we do it in balance.

Involution and Evolution of Monads: When the Monads come into life folded in their 'seed', the events that follow are connected with their purpose and the timeframe suitable for the fulfilment of this purpose. H.P.Blavatsky states in *The Secret Doctrine* that there is a finite number of Monads coming out of Spirit in any Manvantara (the period of the universe's manifestation), even though their number is astronomically huge!

All the Monads emerge from the undifferentiated condition of the One (Spirit) of Pralaya (the period of rest). Their consciousness depends on the level of development that the Universal Consciousness has achieved in the previous Manvantara of the universe's manifestation because, according to H.P.Blavatsky, "cosmic ideation ceases during Pralaya".[3]

Light provides the Monads (the units of consciousness) with endless possibilities for reflection and creation through the manifestation of holograms. As consciousness is developing and evolving so the manifestations are becoming more sophisticated. The development of consciousness throughout all the kingdoms (Mineral, Vegetable, Animal and Human) of Nature is leading to the development of life in all its diversity. The timeframe of consciousness unfolding as a common garden flower is relatively short, but the timeframe of the involution and evolution of humans is much longer because our inherent purpose requires much more life experience.

After *Kalagia, we* call human Monad the 'reactor of evolution',[4] in which the Fire of consciousness is unfolding its petals one by one. Human Evolution – the unfolding of the higher principles and planes of human consciousness – is going this same way. Whilst humans develop as observers, by reaching within and perceiving the archetypes at hand for their development, they are gradually transmuting their bodies and all the qualities of the Solid World that they need for the fulfilment of the purpose of their lifestream.

The Spirals of Involution and Evolution: The processes of 'involution' and 'evolution' of a Monad go in cycles. Every consecutive cycle brings the Monad some new experience, and as result some new quality. Time, the inner space of Matter, is a spiral formed by the cosmic rhythm. Development with Time therefore also goes in spirals. So, the archetype of 'involution' and 'evolution' is a double helix – descending into 'involution' and ascending towards 'evolution'.

At the beginning of manifestation, a Monad descends into 'involution' (materialization), to receive its experiences, go through them and make something new and valuable for Spirit out of these experiences. At a certain stage of development, consciousness changes its direction of development to 'evolution' and turns its carrier into an ascending being.

There may be an individual helix of development for a Monad and a collective helix followed by a large group of individuals. In the *Occult Glossary* by G. De Purucker we read: "mankind as a Life-wave passes through seven evolutionary stages called Root-Races."[5] Every Root-Race of humanity belongs to a certain coil of the collective helix of humanity's development. Every Root-Race inherits some qualities of the preceding Races and lays the foundation for its descendants.

As H.P.Blavatsky expresses it in *Isis Unveiled*:

> *The germ of the present human race must have pre-existed in the parent of this race, as the seed, in which lies hidden the flower of next summer, was developed in the capsule of its parent-flower; the parent may be but slightly different, but it still differs from its future progeny.*[6]

H.P.Blavatsky shows the development of human Root-Races in a diagram in *The Secret Doctrine*.[7] Blavatsky sees the currently dominant 5th Root-Race of humanity as a race that should bring its spirituality into harmony with its highly developed intellect. This is the condition for the manifestation of the upcoming more spiritually inclined and developed 6th Root-Race on the Earth. So, now is the right time to turn the direction of movement of humanity's consciousness from the descending helix of 'involution' (materialization) upwards, into the ascending helix of its 'evolution'.

The Rules of Cosmic Harmony: Cosmic Laws govern the cycles of cosmic development, and the cosmic rhythm makes the motion of 'involution' and 'evolution' orderly and harmonious. The same laws of life, if obeyed, make human life and development harmonious. It is the responsibility of anyone of us to obey the cosmic laws and to fulfil their requirements. *Agni Yoga* asks:

> *...can evolution ever be inharmonious? And how can harmony be created without evolution? Yet people prefer to understand harmony as an immobile and inactive state, and use it as an excuse for their irresponsibility. While the world is in convulsion people prefer to sit in sweet oblivion, and call their benumbed condition by the lofty word 'harmony'".*[8]

We cannot conserve forever the state of our personal wellbeing and that of our society, as much as we would like to. If a 'standing wave' of life stops evolving, it disappears. Life undergoes changes anyway, and our choices can only define whether the changes go on harmoniously or not. As *Agni Yoga* explains:

> *However, the strong chords of harmony are tensed, and as they expand they contribute to evolution. Similarly, the benefit of evolution is in its continuously accelerating motion. It is beautiful to realize that the world is perpetually evolving, in ever-increasing striving and motion. People do not look into the future, but secretly dream about an impossible static condition. But there can be no static evolution. The spiral of evolution is an eternal ascent that even the convulsions of chaos cannot arrest.*[9]

The harmony of evolution, as for a human being, as for a cosmic body, obeys certain laws. For example, the process of transition into a completely new quality goes in octaves similar to the musical scale. It has seven steps (development of seven chakras, seven envelopes of the 'reactor of evolution', seven Root-Races of humanity), and then makes a leap into another octave of vibrations – another quality of consciousness – and appears on a different plane of life, another 'globe' as H.P.Blavatsky calls it.

It can be another more spiritually evolved planet. It can also be a developmental leap of the same planet because the planets are also undergoing their development in conformity with the same cosmic law.

As the cosmic Law of Analogy states, the same universal laws and mechanisms maintain order in every part and aspect of the universe – "As above so below".[10] And the symphony of Life obeys the same rules of order and harmony as the life of every part of it. As, for example, the double helix of 'involution' and 'evolution' is the common archetype for the genesis of life, as in the macro- as in the micro-world.

Evolution of the Universe. Gravitation. Black Holes: *Kalagia* says: "Consciousness is Gravitation creating Life by Thought, by Waves in Thinking".[11] So it appears that according to *Kalagia* gravitational waves actually are Waves of Thought.

Gravitational waves were predicted by Einstein in 1916 on the basis of his theory of general relativity and later discovered in the cosmic events connected with the so-called 'Black Holes'. These waves carry energy of gravitational radiation, in fact the energy of Thought. This energy is similar to the electromagnetic energy of Light and is needed as its counterpart for the formation of the interference pattern in the holographic process of manifestation (see fig.1). The waves of Thought actually form the 'working ray' and in this way participate in the creation of Life.

These Waves of Thought from the Source of the Universal Consciousness possess the power of creating (manifesting) Life through the medium of Light and of unifying the subjects of Life. Universal Consciousness/Gravitation imparts the power of attraction to all the subjects of Life in which It abides.

Agni Yoga calls this power of attraction the 'Cosmic Magnet'. In conformity with the Law of the 'Cosmic Magnet', the cosmic Essences of stars and planets also have their cosmic magnetic powers inherent in the 'seed' of every cosmic Essence.[12] They can also send waves of their thought, their gravitational waves, in conformity with their consciousness. They are co-creators with the Creator in the cosmic theatre of life.

After their manifestation, all the planets and stars in the universe evolve in accordance with cosmic laws. All of them have their Monads (units of consciousness) to develop. All of them change their vibrations and gradually rise into the higher planes of existence.

Astronomers have observed the dramatic transformations and even 'death' of stars. The final, most dramatic, event seems to be when a large-mass star disappears in a so-called 'Black Hole'. If we change our observer's point of view, then instead of seeing a Black Hole as a 'star eater', we will start seeing the attraction of the 'Cosmic Magnet' in action. The formation of a 'Black Hole' may happen when a large star has achieved complete unification with 'All That Is' in Spirit.

In this case, when the Manvantara (the period of manifestation) for the universe comes to its end, the Solid World of the universe disappears in a 'Black Hole' turning all of its remaining matter into energy. 'Dark energy' because its 'Light energy' doesn't spread. The universe is going into the state of 'Pralaya' (literally 'dissolving'). The cycle, which we started describing in "The Urge to Live", is complete.

So, we have closed the great cycle of Creation and Evolution. And yet, from the standpoint of Spirit's development, and that of Universal Consciousness, this was not quite a cycle but rather a helix, with its leap forward. Why? Because Universal Consciousness has developed further 'awareness' through the cycle of Its manifestation as a Universe. We can also say that we, the creators and co-creators of Spirit, all of us together, in the process of our experiences, choices and learning, acquired new qualities of consciousness during our lives, which will show themselves in the new lifetime of our future 'Manvantara'.

The Observer

Let's now focus on the role of the Observer as the Co-creator of the Creator. Let's discuss the difficulties and opportunities of observation, the possible results for a human observer and their application in our daily life.

The Creator and Co-creators as Observers: In the Fiery World, the World of the active Creator's Thought, the Logos manifests the first hologram as the Observer, the mediator between the Creating Spirit and Its creation. The Creating Spirit expresses through the Logos the potential for the archetypes, laws, rules and mechanisms that define the manifestations of life. In the Fine (Subtle) World, the Ideas of the Creator are represented as focuses of energy – crystals of Light – observed by the enlightened beings of the Spiritual Hierarchy. The manifestation of this great potential is completed in the Solid World, by the inhabitants of this World, to the extent of their abilities and particularities as Observers.

We have Observers everywhere in the universe, on any level of life. This is so from atoms, molecules and cells, up to planets, stars and the universe itself. We all observe and co-create life. When observers – participants – collaborate with each other, they can achieve a harmony in what they create together. For example, our organism collaborates with millions of bacteria that abide in our gut and other organs, where our organs and bacteria assist each other.

Contrariwise, when it is difficult to bring different observers on different levels to the state of cooperation, they may bring conflict into the process of creation. In this way the balance of life becomes disturbed. Perhaps this perspective provides a valid point of view from which we can understand the progress of disease within our bodies and disturbances between ourselves and other observers.

Let's discuss our human abilities and particularities of vision as Observers manifesting the potential we are given by the Creator. In our current Monad's spiritual evolution over countless ages, we have 'observed' with the 'eyes' of beings from different kingdoms: as a rock, as a plant, as an animal, and now as a human being. The human embryo carries memories, and even physical resemblances, to the archetypes that our Monad (the unit of consciousness) manifested in its different past incarnations, including the incarnations in other, lesser, kingdoms of life.

Our vision has been educated and conditioned by the experience of observation and manifestation over the multitude of lifetimes of our lifestream. As a result of this experience, humans are manifestations of a much more sophisticated Monad. We are relatively self-conscious observers who can manage and change how we observe and what we manifest by consciously applying our willpower.

All of us are Observers and the holograms that we manifest together are creating life as we see it. Each of us manifests our own life, and it very much depends on our understanding and assertion of the picture of life as we are accustomed to seeing it.

Holographic Vision and its Applications: Holographic vision sees simultaneous manifestations of any event and phenomenon in as many worlds and planes of existence as consciousness is able to perceive. Being the Observers – co-creators of the Creator – all of us have the ability of holographic vision as our potential. In the process of evolution – unfoldment of what is within – this potential unfolds to replace our limited vision of everything happening as a collection of separate unconnected events and phenomena with the holographic vision of deep connections that unite everything into a multi-aspect picture of life.

When seeing holographically, we perceive the signs and the archetype of the whole picture in every particular situation.

We go from the particular to general, and back again with ease because we perceive the unity of creation expressed in the Law of Analogy ('As above so below') and in the property of a hologram, where the whole is reflected in every part, in action in the holographic picture of the world.

Holographic vision grants us immense possibilities and the advantages of clear understanding. Real synthesis of information from different sources, of knowledge and experience, is impossible without it. When a scientific statement or a theory is introduced for our verification we can apply our notion of the properties of a hologram where everything manifesting in a part should also manifest in the whole hologram and in all of its parts.

The Distraction of Manipulation: In our time, the greatest distraction to our free and unobstructed observer's approach is the manipulation by someone (or a group) who has some sort of desire, most often power lust. But the manipulator's agenda is hidden from our eyes. Manipulation feeds on our own desires and fears and subtly suggests its perverse solutions.

The advantage we always retain as human observers is our freedom of choice. When we are presented with information we can choose our subsequent actions. The main danger of manipulation is that the manipulator chooses and structures the information which will be given to us according to their own hidden agenda. As a rule, the observer is given a false agenda adjusted to his or her supposed desires and fears. The observer, unaware that he or she is presented with a warped picture, makes their choice. Such a choice is usually in the interest of the manipulator and the observer will later face the unexpected consequences.

The 'virus' of manipulation is right now infecting the masses via mass communication: the internet, TV, radio, video games and newspapers. However, paradoxically, the same means of communication also provide us with a wide range of information which could protect us from being manipulated. This protection is working for us only if we are trying to deepen and widen the horizons of our observation. In this case our holographic vision is able to show us clearer pictures of reality and not a limited picture tailored to the needs of the manipulator. It immediately opens before us vistas of life related to the subject under discussion, with a wider range of facets than before. We can therefore derive our conclusions and make our decisions from this expanded picture.

Holographic vision can help us to verify the truthfulness of a person and media outlet. When we notice even small lies and inconsistencies in their behaviour and the information that they provide to us, we should ask questions about the truthfulness of the intentions of potential manipulators. Do they 'practice' what they 'preach' to others? We should teach our children to open their holographic vision and to apply it in their life. They should follow the habit of looking for a wider picture and the connections between its parts. This will also help them to escape the trap of manipulation.

Different Angles of Vision: We usually look at everything in life from our personal vantage point. This perspective may or may not coincide with the perspective of those with whom we communicate. It often leads to misunderstandings with the people who observe life from a completely different angle.

A person's angle of vision depends on their background, including such factors as heritage, family upbringing, education, common understanding and sensitivities of a person's social environment, and especially personal experience, in both the current and former lifetimes. This is a very conditional and complicated aspect of anyone's personality. It usually undergoes changes under the influence of the changing circumstances of life and of other people we may meet in the course of daily life.

Our personal perspectives have a great influence on our outlook, choices, and appreciations. For example, there can be an individual who appreciates immutable order and structure in everything he sees, and another who sees structure as a mental prison. There can be a natural leader and another person who feels uncomfortable without being a follower. People are of all different kinds: with rather negative or positive outlooks on life; anxious and careless people; judging and forgiving ones. And so on, in every possibly 'shade'.

However, what unites us is that we are conscious, sentient beings. This provides us with the opportunity of gaining knowledge, expanding our angle of vision and evolving our attitudes and assertions. It enables us to start looking from another viewpoint and in this way changing our whole outlook of life. By changing our perspective, we are able to positively influence not only our personal wellbeing, but the whole social climate. We can manifest a completely new hologram of life for ourselves and have a positive influence on others.

On the contrary, when a person or a group of people becomes subject to manipulation, when, as a rule, their angle of vision narrows under the manipulator's influence, their altered perspective can have dramatic consequences. We can see historical examples of this process in the events of our time, some of which amount to genocide.

There are various ways and methods of altering the angle of observation. For example, consulting psychologists using Cognitive Behavioral Therapy (CBT) are doing just this – teaching their patients how to change their angle of observation and start developing positive plans for the future. However, nothing can change if a person doesn't put in their own effort and willpower. And nothing enables a person to change their angle of vision better than experience (positive or negative). It can have a dramatic effect in either case.

Obviously, for a social leader, certain qualities of an observer, and especially the angle of vision, are particularly important. They eventually define the outcome for the society led by such leader.

For example, an observer who sees the emerging new generation of young people as inherently lazy and troublesome individuals will see restrictions and deprivation as a best way of handling the situation and will support actions of this kind. On the contrary, an observer who sees endless potential and goodwill in the same young people will choose a stimulating and supportive approach, albeit with the necessary discipline when needed. And this difference in observation, if it's put into action, may affect the outcome for the whole future of the society.

That's why developed holographic vision is an absolute 'must' for a leader! This will allow him or her to see and consider different opinions and perspectives. Holographic vision will also enable the leader to foresee possible ramifications of the results of his/her choices. It will open and showcase the full picture of real opportunities and possible deficiencies that his/her decisions may bring to their society, to humanity, and our planet as a whole. Quite the contrary, a narrow-minded leader with a limited vision is a recipe for disaster, both in the present and on into the future.

The Way of Evolution**:** How do we develop our vision and go further along the path of evolution? On the one hand, through acquiring knowledge and experience. Knowledge gives us keys to understanding. When we live through an experience we receive an opportunity to observe different phenomena in close proximity. We observe the connection between the causes and effects and thus we learn a lot. In time this allows us to see the archetypal picture of the potential of an event or phenomenon with a clearer vision.

On the other hand, through aspiration towards lifting our vibrations to the highest vibrations of the Higher Worlds of

Existence and adequately manifesting this potential in everything that we are and do.

For this purpose, *Agni Yoga* advises us to try to emulate the enlightened beings of the Spiritual Hierarchy in all their qualities. In their Love, Tolerance, Selflessness, in their Sacrifice, and in the Purity of their Thought and Intent. By doing so, we call onto ourselves an onslaught of experiences which make us really experience, practice, and express these qualities through self-conquest, as the Masters have done before us. Because the Masters have experienced and conquered all these possible life situations before us, it qualifies them to teach us on the basis of their experiences and shows us that if we persevere we will succeed.

By purifying our mind in this way, we become absorbed in the Light. We become able to see it in and through our mind and life. We start consciously building a body of Light for ourselves. We share this Light with our surroundings both near and far. As the Buddhists say, "The Sage brings the flowers onto bloom as he walks through the garden."

Eventually, as *Kalagia* advises us, we can identify our observer's vision with that of Spirit (or Logos), and our co-creator's consciousness with the Consciousness of the Creator:

> *Do you see the Light and the Fire inside? Act so that Spirit would focus Its Consciousness upon you. By this action you will fix Its Consciousness on you. You will be saturated with Fire and Light like an empty jar fills with the nectar of wisdom; you will feel energy and power, joy and sorrow, and many other feelings simultaneously. When you 'tie' Its Consciousness on you, then you have to act in a way that It would wish to understand you in your completeness. Let It think. By this action Its Consciousness will purify you to such a degree that you would not differ from Its Consciousness and you would be united with It.*[13]

We are Immortal

Life is Endless: Life stems from the Creator's Thought. The Thoughtform resulting from the Creator's Thought gets stretched into a lifestream by the endless spiral of Time. The Matter of a lifestream is manifested holographically, and every hologram is animated by Spirit.

We are inherently immortal and have a lifestream of incarnations in our possession. Indeed, how can a creator die as such a creator carries within the immortal Creator Spirit? In the Creator's Memory where there is no Time and everything co-exists simultaneously, there is no death.

Our consciousness is an eternal observer of the records in the Creator's Memory manifested holographically in the Solid World. One such hologram is the human physical body which is periodically brought into a 'solid corporeal' condition.

Kalagia reveals the meaning of this grand design: "Life is the endless transmutation of death into the growth of Spirit, into the concentration of Its possibilities. This is an aspect of Matter's spirituality – aspiration to the Future, and in this way, the formation of the ingredients of Order from Chaos." [14] So, life is endless. It appears out of the Creator's Thought, aspires to the future, turning chaos into order, contributing to the growth of Spirit, and eventually returns to the Creator.

Human Perception of 'Death': When it happens that a certain hologram is not purposeful anymore, the Creator (creator of the body, Spirit) withdraws Its Ray (the 'Silver Thread') from this physical body. Then this solid form 'dies' and is destined to gradually dissipate.

The hologram of the solid body may dissipate, but in the Creator's Memory the Observer remains alive, and can receive (or rather co-create by Thought with the Creator) another holographic body. This may be either only an energy body for the Fine World or also a body that we perceive as a solid body, for the next incarnation in the Solid World.

We seldom realize consciously during our lifetimes why the Creator's Ray has been withdrawn, unless we are aware of the purpose and task for which we incarnate on Earth. But consciousness in its highest state is aware of it because it takes its awareness from the perfect picture in the Creator's Memory (in Sanskrit – the 'Akashic Record').

If during incarnation we were aware of our highest state of consciousness, it would certainly change our outlook on matters of 'life and death'. We would be able to find the reasons for some events happening right now and their connections with the past or even the future in terms of the Solid World sequential time. We would be able to see actions and their consequences that some Oriental teachings, such as Hinduism and Buddhism, call 'karma' (from 'action, work or deed' in Sanskrit).

As we understand it, there is a cosmic Law of Karma which connects the trinity of causes, effects and lessons of life, not only in one lifetime, but through all the incarnations of the human lifestream. [15] Collecting and balancing karma is an ever ongoing process. 'Life and death' of our physical bodies is an aspect of this process.

So, what happens after 'Death'? Commonly consciousness appears in the plane of the Fine World to which it belongs. In the Fine World it is in a different condition of Time and not restricted by solid body limitations. It is easier to understand everything that happened throughout the incarnation in such state. If consciousness decides to return to the Solid World to continue learning there, it returns with the understanding gained during the 'inter-life', or 'life between lives', and with new tasks to fulfil in the light of this knowledge.

Death is not the end! Rather it is liberation of consciousness from a hologram that is unable to further perform its duties in the life situation in which it appeared. It is also the means for the development of Spirit's self-consciousness in a new arena of experience.

When consciousness feels the urge and readiness to return to Solid World because it has become aware of its purpose, the energy body starts producing a new solid body as its 'space-suit', enabling it to live in the low vibrations of the Solid World.

In some rare cases, an enlightened human being whose karma has been sufficiently balanced, i.e. at least 50% of it has been worked off, according to some spiritual teachings, can achieve Ascension. This means that such a person, while still living in his/her solid body, managed to turn its vibrations into the highest vibrations of Light.

These vibrations can be attracted straight away to the highest Fiery World of existence and unless such a consciousness chooses otherwise, no more incarnations follow in the current Manvantara.

As a matter of free choice, cosmic justice allows a mature Soul, who has passed all the experiences needed to balance karma and to become part of the perfect picture in Spirit's Memory, to incarnate on Earth as a Teacher, a Redeemer, as a so-called 'Boddhisattva'.

Human Options and the Master's Advice: When the physical body dies, the direction in which the departing immortal aspect goes very much depends on the final thoughts of a dying person. These thoughts are made up of a lifetime's habits of thought.

Agni Yoga provides us with plenty of hints and descriptions of the possibilities that await human spirit after it leaves the boundaries of Earth's attraction. Ascended Master M. who dictated the *Agni Yoga* teaching to Helena Roerich encouraged us to aim high and invited us to the higher planes and further into the vast spaces of the Cosmos.

In *Agni Yoga,* we can find the following instructions on the departure of human spirit from the physical plane:

For a consciously developed spirit the period of sojourn on the astral plane could be limited to the interval of forty days, but various earthly conditions have prolonged this time to an interminable period. The misery and grief of those who are carried away from Earth binds them thereto.

The best instance of this is found in the Biblical legend about Lot. For a new life they walked out of the city, and only one condition was imposed upon them – not to look back. But Lot's wife looked back, and bound herself to Earth.

Religion says: He who goes to his fathers will dwell with them; he who goes to the angels will dwell with them; and he who goes to God will dwell with Him. It means that he who has ordained himself the utmost progress arrives at the best attainment. Therefore, the best parting words to the one who departs from Earth will be – 'Hurry, without looking back.'

What about the dear ones? But the higher you ascend, the better and closer you will see them. Of course, the cause of the delay is usually in the last remaining near ones. Therefore, shortening of the sojourn on the astral plane depends upon a proper cooperation.

The higher up the more pleasant the stay; and on the border of the mental plane the spirit can rest, because there the spirit is already subject to lofty attractions. But one must consciously avoid the lower strata. It is necessary that an explosive impulse of the consciousness propel the kernel of the spirit upward as far as possible. Therefore, the moment of transition is so important, for in it one may dispatch oneself to the higher strata. Once the lower strata are contacted, it is very difficult to rise afterwards.

If the condition of spirit permits, it is far better to use the last flash of the nerves' emanation for flight. Thus the lower strata will be sparser. It is important to dissolve the atmosphere of the lower strata so that it will not press upon the Earth. Cooperation from above and below will give the speediest results.

Transition without consciousness has been correctly noted.

It is easy to assist in this, if beforehand there be strengthened the desire for lofty flight. Then the emanation of the nerves acts almost automatically.

Prayers about the 'wandering of the soul' are very helpful. The one thing wrong in them is that they reiterate messages about rest, whereas it would be better to stress haste.

Everyone should read and remember this, for it will not only help the individual but also advance the world plan.[16]

Mortality and Immortality in Human Perception: Our human perception of Life and Death usually depends on where we focus our attention. Time becomes our demanding and unchallenged boss if we are mostly focusing attention on the goals and problems of our earthly life. Commonly we are led through our life experiences by the 'Serpent of Time' accompanied by obsessive fear of death.

But there are increasing numbers of people who already feel themselves as creators. Their lives are devoted to the process of creation. There are spiritually mature people who come into incarnation with knowledge and a strong inner feeling of their purpose. They often have the feeling that they have an aspect of themselves which is immortal. At least, they feel they are creating something which will never die.

There are creative people like Leonardo da Vinci or Mozart who leave after themselves immortal creations. The attention of such people is most often focused on their search for the archetypal pictures, music or ideas. Actually, they are searching for the Creator, if perhaps unconsciously. And there are people who directly search for the Creator and Its Universal Truth.

All of us are created with Rays of Light. This Light flows through us creating for us an opportunity of co-creation. If we concentrate on the workings of Rays, passing them through us and participating in the flow of unrestricted receiving and giving, we become 'givers of Light'. We also become 'givers of Love' because giving is a property of Love. This is a deed of immortality. On the contrary, when a person focuses solely on the mundane aspects of life, then he or she establishes their mortality.

Actions that follow the attachment attract karma. Karma requires balancing through re-incarnation until all of it is worked off. As a result of all the experiences and learning, a human being gains the ability of crystallizing their consciousness to the condition of a pure crystal and the ability of passing Light through it.

Agni Yoga tells us:

As a heavy fate the thought of death hangs over the human consciousness. The specter of death is present as an inevitable cup, and having traversed the entire path of life, the spirit concludes that here one must terminate one's existence. Such is the plodding of the spirit dissociated from Cosmos.

Ignorant of the beginning and seeing only the end, the dissociated spirit aimlessly passes through life. But everyone may earn immortality by admitting Infinity into his consciousness. Fearlessness toward death and striving towards the Infinite will provide the spirit with the direction to the spheres of cosmic endlessness. Affirm yourselves in the acceptance of immortality, and infuse into each of your actions a spark of the creativeness of the Cosmic Fire, and that inexorable fate will be transformed into the one call of cosmic life. Our great, just law has chosen you as participants in the universal manifestations! Cognize immortality and cosmic justice! A beautiful step is prepared for everyone. Find the path of thinking about immortality![17]

Question: "What if a person fears life more than death? This can happen sometimes when a person is in extreme pain or hardship, and commits suicide. Even, God forbid, a suicide terror attack, killing not only themselves, but many other people."

Answer: "If we are still alive, it means that our purpose on Earth is not yet fulfilled to the end. We still have choices to make, difficulties to overcome, and lessons to learn. Life is much more expansive than the 'corner' in which we feel ourselves driven. Fear of life is not the best advisor because it narrows our choices to the simple alternative: to live or to die. Instead, we should ask our Higher Self what we should do to overcome our fear of life. In many cases, unexpected help comes right at the moment when all opportunities seem to be exhausted, but the person is just holding out and not giving up their life.[18]

In our many lifetimes we go through different experiences, until we come to the condition of an Archetype which has perfected knowledge as far as this is possible in this Manvantara. If a person chooses suicide as their experience in the current incarnation, this person should be aware of the karmic consequences of suicide. He or she would be compelled by their own immortal consciousness to return into another incarnation to balance their karma and to eventually fulfil the purpose which they had failed to fulfil in their previous lifetime. So the person returns into a similar situation as many times as it takes to gain new expanded understanding and to make choices different than before, choices which will balance the accumulated karma.

In the case of a simultaneous murder – suicide, the murderer is accountable not only for his or her crime, but also for the victim's unfulfilled purpose. So the murderer removes and undertakes karma, and the task of balancing it, from the victim or victims. As a result, the victim's karmic condition becomes lighter, and in some cases it is even enough for Enlightenment. At the same time, the burden of karma for the former murderer is heavy, the conditions of balancing it are hard, and it may take not one, but several incarnations to clean up his or her karmic debt. That's why Agni Yoga proclaims that it is better to become a victim than a perpetrator."

Question: "What about people, such as Japanese kamikaze pilots in World War II, who are honour bound to commit suicide in some situations? What about people with mental illness? What about people sacrificing themselves to allow others to survive?"

Answer: "We have to understand that suicide is not a crime, but an experience. And we cannot judge from our limited knowledge looking on as outsiders. If a person chooses to go through such an experience, the person's consciousness possesses all the knowledge of this person's considerations. After the death of the body, consciousness will become aware of the full extent of karmic consequences of the suicide and will have the opportunity to choose how to deal with them in the following incarnations.

Anyway, all of us, in multiple incarnations, go through different experiences to learn and to come eventually to the crystalized understanding of an Archetype."

Working with Thought

What is Thought?: Everything starts with Thought and from Thought. Whatever existed in the universe always came into being because of Thought.

Kalagia defines Thought thus: "Thought is a Ray of Fire-Carrying Light which is creating the hologram of that what is Occuring in the Substance of Space."[19] According to *Kalagia*, Thought spreads in the very substance of Time and because of this it spreads instantaneously in Space.

Through the process of holography, Spirit's Thought creates and animates all the forms of Life in the three Worlds of Existence. We therefore call these created phenomena 'thought-forms'. Thought-forms carry the essence, the Idea, of the Thought, and are imbued with the power of Spirit by this Thought.

The creating Spirit lays the foundation for the manifestation of the archetypes of Life through Its Thoughts and their essence – Ideas. Together, these archetypes form the Perfect Picture of

Creation. Human beings are Creator's co-creators, and they also constantly work with thought, especially when they wish to experience or create something new in their lives.

Multitudes of thoughts created by people are always 'flying' around producing elemental thought-forms which belong to the Subtle World. They are invisible to our solid body sensors. When someone 'catches' a thought and it ignites their consciousness, the thought is put into action producing new experiences, creating new holograms. Otherwise, the whirlwind of thought's energy may 'fly' around until its energy dissipates. That's why our thoughts often fail to manifest – for better or worse.

As a human being carries the essence of Spirit the Creator within, some human thoughts can be a direct multi-dimensional reflection of the Creator's Thought. That kind of thought often comes to us when we are in a momentous state of 'here and now' where Time is non-existent, and we appear simultaneously in all the three worlds of existence.

In that state our consciousness is unlimited and reaches the all-dimensionality of Infinity, bringing us revelations of Truth. Our Thought is producing crystal-like archetypes of holograms of Matter. Because of the diversity of our personalized experiences, a consequent diversity of archetypes appears. Their crystals are created in the same way that the first basic archetypes have been created.

Thought carries real energy, a vibration of high frequency, coming from the Fiery world and beginning its journey into manifestation through the consciousness of living beings who possess the ability of perceiving and working with thoughts. Living beings can sense and convey their thoughts by different means. For example, some extra sensitive people can see thoughts as fiery whirlwinds.

The Process of Thinking: Everything originated in Highest Consciousness as an Idea. In the process of thinking, human beings take the thoughts from the universal stream of consciousness and apply them for meditation (thinking) and creation of new thought-forms. This meditation is one of the main purposes of human existence. It is the means of learning and co-creating with Nature.

Very often in our life, the process of working with thought starts from a Question which we have stumbled upon in the course of our life experiences. Willingly or not, we concentrate on the question. The question becomes insistent and whilst trying to find a correct Answer we meet considerable Resistance. Resistance causes Tension, which shows a tendency to rise every time we try to pierce into the substance of the Answer. Circumstances or people may become carriers of resistance and be conductive to the rise of tension. It is the most difficult and demanding stage of the process.

When tension reaches the necessary level, we feel that it is impossible to continue to concentrate any further. Then we expand our consciousness with readiness for the perception of everything that may come. Depending on our developed habits, we may meditate, fall asleep, or immerse ourselves in water, and then we receive the Idea. It comes to us as a crystal of thought, an Answer to our Question, and simultaneously to all interrelated questions.

The Fire of the initial Idea ignites our consciousness and enables it to start its fiery meditation, which engages our fine structures and eventually the brain. According to *Kalagia,* if we already have in our mind a permanent kind of thought – an Idea – then we start attracting a similar kind of thought out of the multitude of circulating thought-forms with our inner magnet. By repeating this process of fiery meditation continuously, we gain the Power of Thought and can apply it for any type of creativity.[20]

We can sometimes observe and participate in the process of 'thought-swarming' of our collective consciousness when likeminded people are working together for a common purpose. They unite in the process of thinking attracting related thoughts and working with energy which is multiplied by their united creative work. Usually we call it 'brain-storming'. The brain is obviously involved in the process – but it is much more than that!

We also come to understand the universality of experience. Experience, be it exoteric or esoteric, scientific or mystical, is at the base of any real knowledge. No divide exists between thought processes of the so-called 'spiritual' and 'material' or 'physical' sciences. They differ only by the frequencies of vibrations they are dealing with in their experiments (experiences), principles and the languages they use, and rules to which they adhere to arrive at conclusions. We can also say that they differ by the World in which their tools work: Solid, Fine or Fiery. However, to make a real discovery, you have to apply the Power of Thought, and this is an attribute of the Fiery World. Your consciousness has to be really in tune with the wavelengths of the Fiery World to be able to perceive the Idea.

Putting Thought into Action: Let's scrutinize the very process of making our Thought work, create and manifest. According to *Kalagia,*[21] our steps start from Attention which we direct towards the object of our attention, even if our object is ourselves. Attention is followed by Perception. Perception comes as a result of the cooperation of Fire and Matter, and consists of four stages: sensation, impression, notion, and imagination. Together, these bring us fiery understanding – awareness – of what has been perceived. Any living being, as an Observer, Spirit's co-creator, comes to awareness of what they perceived in their own way. Even plants and animals somehow go through this process.

However, different kinds and levels of awareness lead to different results. The awareness of an artist looks different from the one of a scientist, and still different is the awareness of a musician. Also, the deeper awareness, the higher the creative abilities of a person.

To receive awareness is not enough for action to begin. This is only the start of the process, and now a question appears: how can we put our awareness into action? *Kalagia* advises us to find a formula for the form of our awareness. Formulas can also differ for people of different mindsets and ways of expressing themselves: for mystics, scientists, poets, musicians, artists and others.

As an example of putting Thought into action, *Kalagia* suggests to use a word as a formula. By verbally defining our object we finalize the process of perception. We actually bring the priceless treasure of the Idea from the Fiery to the Solid World where we can deal with it from the standpoint of logic.

Verbal logic is a part of the process of thinking. Thanks to our Solid World direct experiences and scientific research, we are used to making logical conclusions. However, according to *Kalagia*, verbal logic deals only with the outward form of a word. The energy of the essence of a word can be applied only by switching on imaginative thinking. The energy of thinking returns in a cycle from the word back to the awareness of its contents by stages: imagination, notion, impression, and sensation.[22]

Thus a switch is activated from verbal logic to images. If we successfully switch from verbal-logic thinking to imaginative thinking, then we are in control of our energy. We are penetrating into the might of the inner powers which are the cause and the source of everything manifested. We release these powers outward and they produce a wave of Time density. The pressure of this wave is immense. It can destroy or create depending on the Thought. This can be the climax of co-creation (manifestation) when the Thought gets energy enough to make its 'working ray' actually work! Call it 'mystic', or call it 'genius', or simply 'the mechanism of creation by Thought'. Whatever you call it, this is a crucial phase of the process because the inner powers are connected with the actual content of the object and by our attention can be put into practical use.

So it appears that any art or science, when making a real discovery, goes through this mystic penetration into the depths of the inner powers. The difference is perhaps only that an artist has no need to go through the stage of verbal logic.

Like *Kalagia, Agni Yoga* also stresses that human thought governs the mighty powers of the fiery energy of the Source – 'Primal Energy' – as it is called in *Agni Yoga.* There is a great responsibility which humans bear in this process:

> *It should be recognized that the energy of thought is one of the highest manifestations of Primal Energy. It is impossible to isolate thought from the fundamental energy of Cosmos. Precisely, it is thought that eternally moves the fundamental energy. Thought generates certain currents that serve as the awakeners or, so to speak, the rejuvenators of the Universe. Thus, when I say that thinking beings participate in world-creation, it can be understood literally, not allegorically. It follows then, that man has a great responsibility for the quality of his thought. Each kind and strong thought produces beautiful vibrations, but an evil one strews Earth with deadly dross.*[23]

As we can appreciate, the future of our planet and its inhabitants depends a great deal on what we are thinking right now!

Experience of Thought Materialization: As we already discussed in "Manifestation on the Boundary of Worlds", a group of Russian prisoners who called themselves the X7, experimented in an underground Siberian salt mine with manifesting their Thought as Creative Word in a material form.[24] They observed the very process of manifestation: from the appearance of a faint impression of their thought-forms as they were infused with the power of Light, to the manifestation of real forms of Matter.

The X7 mention the Solar Substance as the source of power. Solar power represents for us, the inhabitants of our Solar System, the Fire and Light of the Creator. The Essence of the Spiritual Sun behind the sun that we see with our physical eyes is the source of this Solar Substance. It possesses immense Power which sustains and nurtures our Earth and the other planets of our Solar System and provides vital conditions for the life of all of us.[25]

The X7 write about their experiences of applying the power of thought for the manifestation of forms of matter out of Solar Substance: "We then were able to see the necessity of the long ordeal of erasing all that was unlike the pure essence of Solar Substance. It could only be conceived when met on its own level of being. It became malleable and responsive to thought only when that thought equalized with its inherent awareness of pure being. We could not use it to create our desires in form, unless those desires were as clear and pure in essence as were the radiations from which they came".[26]

So the longest and most difficult part of working with thought is to purify our mind and raise the quality of our thoughts to the level of archetypes, where our idea can find energy for its realization. When our mind is immersed in mundane thoughts and closed to the higher radiations it is "warped and limited in its power to receive". "But each degree of expansion in the consciousness of those who bear the Light elevates the whole stream of consciousness."[27] In this way we can cooperate with universal consciousness and participate in the elevation of humanity's consciousness.

On the Path of Evolution: Raising our Thoughts to the level of archetypes actually means reaching into the depths of our essence. We are constantly searching in the world around us for solutions to our inner questions and problems. There we can find a lot of useful information. We sometimes receive crystals of knowledge from the Masters, books, earthly teachers and each other, but only soul work can find the right way for each person – the designated path of Evolution.

In the process of Evolution (return to Spirit), which is opposite of Involution (Materialization), we are acquiring the ability to unwind envelopes of Time. This is the Path to the greatest Power – the power over Time – 'Kalagia'. This is the Path to our freedom from the fetters of Matter, as no freedom is possible while our consciousness remains attached only to its material manifestations.

Everything that happens to us in our earthly life is connected with a clear Thought or Thoughts that came to us at certain moments of our life. We are seldom able to specifically notice it because the moment of Thought's appearance and the moment of its realization may be divided from each other by a considerable amount of time (sometimes by decades). It is a result of Nature's way of progression reflected in our lives, where every step into the unknown is connected with errors and lessons. Our faltering steps along the Path of Life result in the necessity of balancing personal, family, national and even global karma, as we see today for example, with the global warming environmental crisis.

While we are spiritually developing and balancing our karma, our Thought starts attracting its realization much sooner than before. This means that we are gaining power over Time.

When a person is ready for Liberation, he or she comes into incarnation as a strong magnet attracting all the remnants of karma and all the possibilities of balancing it. But this situation doesn't promise an easy life. Conscious sacrifice is usually the only possible way towards Liberation.

Chapter 5

The Ways of Shaping the Future

Ways to the Perfect Design

The Perfect Design or Our Own Doing?: The perfect design of the Creator – what does that mean exactly? Does this mean that everything happening on our Earth has been foreseen and designed in the mind of the creating Source in a perfect way and nothing can be altered whatsoever? Or are we supposed to be active co-creators of the world in which we live?

These are some of the major questions which human mind and soul have pondered throughout the ages. This 'work of the ages' actually becomes a training ground for growing consciousness. The way we respond to such major philosophic questions very much depends on the knowledge and skill that we have accumulated in this and previous lifetimes.

Now, as we have already discussed the manner in which the perfect archetypal design is working, we have some grounds to analyze such enigmas. The perfect design is envisioned by the Creator and is embedded in Its Memory as an archetype, a perfect potential. It is recorded in the highest frequency vibrations of the Fiery World and can be manifested as a hologram in the Solid World in different ways.

We can see an example of the connection between the potential and its realization in the genetic code and its manifestation in an actual human life. The genotype provides the potential but doesn't completely define human life as every faculty and ability can be developed in different ways and lead to quite different outcomes. Similarly, though there is a perfect pattern of Beauty in the Creator's Memory, nevertheless we have created our world as it is, and we are not always happy with how our creativity has turned out! Fortunately, we are given Time to experience, to learn, and to make changes.

Interaction of Worlds: Crystallized Thoughts which make up the substance of the Fine (Subtle) World contain a vast spectrum of ideas. The expressions of Thoughts coming from the Fiery World are transmitted to the state of Fine World by Light. However, the lower the strata of the Fine (Subtle) World the lower its vibrations.

Do we have the ability to reach the Fine World and if so, how do we do it? Yes, we have an inherent connection with the Fine World. Actually, the Fine World vibrations constitute a great part of our energy body. Our consciousness sometimes in sleep, in deep meditation, or through NDE (Near Death Experience), visits the layers of this world to which our attention becomes attracted.

For example, people who have experienced an NDE, often describe a picture of bright, though not blinding light, fields full of beautiful flowers, pleasing music, welcoming friends and relatives who have already passed away, and a feeling of peace and love. Everything is in harmonious movement and the visitors themselves can choose either to be guided further into the after-life or to return to their lives in the solid bodies left behind.

Sometimes people who are experiencing an NDE see pictures resembling the descriptions of hell in Dante's *The Divine Comedy.* Devastated surroundings often covered with ash, and shadowy figures repeating the same pointless actions again and again.

Pim van Lommel, a cardiologist and NDE researcher, in his book *Consciousness Beyond Life,* refers to the memories of one of his subjects who experienced an NDE and describes that what he saw:

The plain was crowded, even jammed with hordes of ghostly

> *discarnate beings... And they were the most frustrated, the angriest, the most completely miserable beings I had ever laid eyes on...Everywhere people were locked in what looked like fights to the death, writhing, punching, gouging...These creatures seemed to be locked into habits of mind and emotion, into hatred, lust, destructive thought patterns..."*[1]

We can presume that these thought patterns of conflicts and battles are recorded in vibrations of some strata of our Fine bodies. Probably these low vibrations of the Subtle World affect our consciousness and make us fight for power in our Solid World. Could it be that unending wars and crimes where people are searching for power for the sake of power are projected by such destructive thought patterns? Could it be that by putting these patterns into motion, we are turning whole regions of our planet into a 'hell'? Shouldn't we better try to find the way to replace these patterns in our thoughts with those of peace, love, compassion and cooperation to make it a living paradise?

Understanding as a Result of Experience and Learning: Why do destructive patterns of thought appear at all? Why do they sometimes appeal to us? Is there any justification for fighting, or is conflict always inadmissible?

To answer this question, we have to return to the very beginning of the lifestream of our Monad's incarnations and its experiences in the Solid World. We need these experiences because we learn through acting and facing the consequences of our choices and actions. Among other things, the experience of fighting teaches us how to discriminate war from peace and how to come to peace and mutual cooperation on the way to the unity of the perfect design.

If we refuse to learn lessons from our experience, the situation then repeats. If we continue acting in the same way as before, we may receive even stronger signals of the necessity to change. If we have learned our lesson and changed, then we change not only our destiny but clear the thought environment around us for the present and into the future.[2]

However, if we are conservative to the point we reject any possibility of learning, changing, or altering our narrow point of view, then we are continuously projecting our inalterable habits of thought into the environment which we are creating with our thoughts and actions. In this way we are creating a 'hell' where nothing ever changes on the Earth and in the fields of thought of the Fine World.

On the contrary, if we come to sincerely regret and repent our past thoughts and actions, which we start seeing as undesirable, and are determined to change, it means that we have learned our lesson and opened the way into a better future. It can be a personal or collective lesson alike; in any case, a well learned lesson expands consciousness and leads to major changes in the thought patterns and in the manifested patterns of life.

Generally, our experiences teach us to connect our thoughts and actions with the consequences of their realization. We 'sow' our thoughts in abundance in the Fine World and also manifest our own and someone's 'seeds' of thought in the Solid World. We 'sow' our deeds in the Solid World and in time reap their fruits. The more distinctly we observe this process the more we understand sowing and reaping as one interconnected phenomenon, and our vision needs less time to go through an experience and receive a lesson. Our vision becomes that of an observer in the position of the 'higher ego' on the boundary of the Fiery and Fine Worlds, where the observer perceives sowing and reaping simultaneously. Evolution of consciousness is achieved in this way.

The experiences through which evolution of consciousness is achieved aren't the easy ones, and some of them dearly cost humanity. The continuing wars show that we humans are still ready to sort out our differences through fighting. That we tend to think that we are always right and don't need to regret or change our actions, especially whilst we are victorious; and our collective mind is busy with searching for power rather than with searching for ways towards the common good.

However, life cannot stagnate; it has to change, to progress. Only by searching for new ways of thinking and means for development can we expand our consciousness, gain awareness, resist manipulation, and come to a more peaceful and harmonious future for humanity.

When we experience and learn, we receive new knowledge and understanding. This understanding appears in all our energy bodies and then we can share it with everyone who asks a question. This inevitably follows, no matter how the question has been asked – directly or not. Our new understanding becomes integrated in the common field of thought. So the answer can be received through any consciousness connected with this field.

The common field of thought is not restricted by any limitations of time and space. A thought that came to us from millennia back in history may carry an answer to the most important and seemingly unsolvable problems of today. This answer will sound anew for the new generations.

For example, Plato said the following words of wisdom: "For a man to conquer himself is the first and noblest of all victories". But what does it mean for a man to conquer himself? What exactly should he conquer?

For a man or a woman it is a twofold task: first they have to conquer their powerlessness, and then they have to conquer their power. By conquering their powerlessness a man or a woman triumphs over their weaknesses and comes to power. Then they should conquer this power, become its masters, and use this power for the tasks at hand. If a man or a woman hasn't conquered their power, they then succumb to this power and become its slaves – hostages to their own power.

Balancing Power: What Plato said is correct for an individual and for humanity. Thanks to our gifts of intellect and intuition, thanks to our inborn creativity, enthusiasm and relentless labour, our civilization has accumulated great powers. Thanks to these powers we have managed to greatly improve our living conditions, create powerful technologies and to extend the longevity of the average person.

However, can we be sure that we have already conquered our power and became its masters? Why then is humanity struggling under the yoke of endless war and destruction? Why do we waste our precious resources on weapons of all kinds instead of extinguishing the poverty and the powerlessness of millions? Why do we allow and accept the suffering of mothers and children which will negatively affect the future of our civilization?

At the moment our civilization is ruled mostly by Power accumulated by any possible means: the Power of Money, the Power of Weapons and the Power of Manipulation. The common aspiration of humanity seems to be accumulating Power and not conquering Power by balancing it. Most probably, because of the unbalanced ways in which Power is accumulated and distributed, it gets distorted and attracts distorted patterns of thought from the Subtle World, bringing us all the maladies, wars and injustice which often horrify us.

As we can see in the modern world, a minority of humanity has already accumulated and continues accumulating more and more Power. Another part of humanity appears in a situation of powerlessness which brings them a variety of suffering. World renowned epidemiologist Professor Sir Michael Marmot, Chair of the Commission on Social Determinants of Health, repeatedly stated in his public lectures that poverty and powerlessness may cause depression and related mental and physical health problems. This situation doesn't stop with individuals. It causes social calamities of different kinds on a massive scale.

Our pursuit of power continues to be a brutish struggle for power over other human beings. We willingly accept a situation where we are 'lucky enough' to possess power over others at the expense of their powerlessness. We behave in such a way until we ourselves, facing the bitter consequences of the law of karma, appear in the situation of powerlessness. Then we start suffering and looking for ways to increase our Power. This vicious circle will continue till we learn to balance our Power.

In "Co-creation with Nature", we spoke about the universal mechanism which governs by harmonizing the Trinity of Love, Wisdom and Power. Applied to a human being, this mechanism requires our Power to be in exact balance with the level of our Love and Wisdom. In this case, we become a harmonious part of an uninterrupted 'flow' of energy which we expend wisely and lovingly.

It is said in the book *Saint Germain on Alchemy: Formulas for Self-Transformation* :

> *In memory's storehouse, the writings of all past actions in present and previous embodiments stand as an electronic record of considerable weight and influence. The Power of this record, together with the momentums generated by the misuse of energy, is a subtle pressure that affects the present moment for good or for ill (relatively speaking) unless it is brought under the control of the balancing Power of the great God flame within the heart – the immortal flame of Power, Wisdom and Love.*[3]

We always receive the Power we need to reach and express Love and Wisdom to the extent of which we are capable. However, whatever our intentions, if by any means we gain additional power that is not in balance with our Love and Wisdom, this power can become stagnant and destructive. This is our human responsibility – to balance our Trinity of Power, Love, and Wisdom. Simultaneously with our efforts to balance our power, we start preferring the power of cooperation, of ideas, knowledge, and skills, which we carry and refine for incarnations, to the transient power of money and weapons. This means that our consciousness has ceased focusing on its 'desire mode' and instead moved to the 'compassion mode'.

A life of joyous, sincere and devoted learning, balance, and sacrifice inevitably raises a person's frequency of vibrations. By natural magnetism, such a person becomes attracted to the vibrations to which he or she belongs. Simply put, such an advanced person is no longer interested solely in those aspects of life which formerly absorbed them – i.e. money, power, lust, etc.

The person who can change their inner condition appears also in a different outward situation and surroundings. As Observer, such a person ceases to see or even pay attention to people and activities formerly natural for his or her environment. Instead, this person starts seeing and paying attention to completely new objects. In this way people change their experiences and the hologram of life for themselves and by example, for those around them.

Directing our Movement towards the Perfect Design. Initiations: The word 'initiation' stems from the Latin 'initiate' – to begin. We can understand initiation as a new beginning through personal transformation. Usually such transformation happens as a leap forward in our consciousness, a revelation after trial and effort to overcome obstacles and conquer ourselves.

As theosophist G.de Purucker writes on initiations in his book *Fountain Source of Occultism*: "It is in fact a hastening of the evolutionary process: not in the sense of omitting any stage, but in condensing within a short period what in the natural course would take eons of striving to attain".[4] We can add to it that hastening the evolutionary process also means lifting the frequency of vibrations. This would be a direct result of any initiation.

At different times in human history, from tribal to medieval, as well as in the later occult societies, initiations have been connected with special rites and rituals aimed at putting a person into the conditions where his or her transformation could possibly be initiated. However, initiation is first and foremost an inner process. Life experiences along the spiritual Path provide an evolving person with possibilities for multiple initiations.

Buddha advised his followers not to look for easy ways in life, but rather to choose the difficult ones because trials on the Path of service provide the best environment for initiations. However, even a difficult path can become enjoyable if you feel it to be your Path, freely chosen, and not forced upon you.

We discussed earlier in the book how differentiation of the One into planes of the manifested universe is going down the scale of vibrations by octaves like the descending scale on a keyboard (see "Manifestation on the Boundary of Worlds"). The same 'stairway' of vibrations can serve us for the return to the Oneness through gradually increasing the frequencies of the vibrations of our energy bodies by octaves.

At a certain stage of development, the frequency of our vibrations raises to a point where they already belong to a higher Octave of vibrations. We then can 'ascend' into this new Octave. This is explained by St. Germain in his book *Earth's Birth Changes* channelled through Azena Ramanda: "For ascension is merely a change in frequency".[5]

Weaving, War, and Wisdom

Goddess Neith. Weaving of War and Peace: Why was Goddess Neith, who was a great mother goddess and one of the most ancient goddesses of Egypt, also called a goddess of Weaving, Wisdom and War?

The attributes of Weaving, Wisdom and War are probably

associated with Neith because of her way of creating through 'weaving' (see "The Birth of Archetypes"). The basis of 'weaving' of archetypes of the universe is provided by the Trinity: Spirit in a state of Unity, and two opposites representing any two opposite forces, such as Light and Shadow, Positive and Negative, and so on. The opposites behave as separate forces, but they are attracted to their unity by the Cosmic Magnet of the core ('seed') of Spirit.

In a Trinity, any manifested conscious beings representing such forces feel the repulsion of division and the attraction towards unity. As two parts of the same Whole they reflect each other in thinking and acting. They are similar in their abilities yet distinct in their qualities. Being two parts of the Whole, they may understand that they are in fact united and work together, or they may focus on their differences and perceive each other as completely separate beings. However they feel, in the Trinity their combined action is 'weaving' the world.

They can do it in two different ways, depending on how they understand themselves. They can seek and reach their common Source through their continuous work of balancing power which they both receive from Spirit. In this case, they would 'weave' Peace. Or they could each try to appropriate the power of the core and make it their exclusive property – this would lead to 'weaving' War.

Seeing differences is the most tempting and simplistic. However, only by acquiring wisdom can they come to understand that they are actually united in the core and the same Spirit is in both of them. To see and think beyond the visibly obvious – this is the path of Wisdom. Goddess Neith represents the Wisdom of Unity or the War of Separation because the 'opposites' can choose either way.

War or Peace: As we can see, the basis of any war is the emphasis on our separateness and readiness to gain power to the detriment of others. Throughout the history of humanity, power has been associated with possession of land and its resources and with power over other people.

Wars for possession of a plot of Earth and its resources continue endlessly. The necessity of war is usually justified by historical, religious or other similarly 'valid' reasons. However, power lust usually underlies all the noble explanations politicians have concocted for the justification of war. Leaders of the opposite groups use the fear and insecurity of ordinary people to gain their loyalty. They encourage people not to peacefully coexist but rather to take sides and to search for safety through belonging to one of the opposites.

In such a way, fears and conjectures of the human mind lead to war. Wars will continue until the consciousness of the majority of people in the world undergoes some cardinal changes. Multiple examples show us that people can live in peace when they are striving for unity yet respecting their differences.

We are united in Spirit. Therefore, trying to pull our unity apart is absolutely pointless. *Agni Yoga* proclaims that we possess only what we create (when co-creating with Spirit) and whilst we create. If all humanity comes to such an understanding then wars will come to an end. Even more, after a couple of generations of evolution of consciousness, people may habitually respect the rights of the planet Earth and all its living beings. In just such a way, now in the 21st century, through many painful challenges, we are coming to respect the rights of all human beings, despite their differences.

The ancient images of the Egyptian Goddess Neith are often associated with archery, probably because manifestation is a dynamic condition. It requires energy, action, and aims at a goal. Manifestation (weaving the world into being) is a process that requires fine balancing of the attributes of Unity and Division. Only Wisdom can hold it in a state of dynamic balance.

Peace is not just the rejection of war. It is the result of

achieving a balance of forces that represent tension between the opposites, which are pulled apart and yet united in their essence. As expressed in *The Mysterious Story of X7*:

> *Peace is always existent, always in exact rhythmic response to the realms in which it is known and accepted. The mind of men must be capable of acceleration to that degree, and as it is stepped up by the opening of the heart centre, peace becomes an accepted condition of being. It is thus that we behold it, but we could not do it until our whole being was aligned with the radiations and we were clear, pure channels for it.*[6]

In many places on the Earth nowadays, we witness wars fought for religious or other similar ideals. The opposites are trying to unite through war in the way they have been accustomed to 'unite' for millennia. People are suffering and looking out for a Messiah who will bring them peace and security.

The archetype of the Messiah is already there as a potential in the Memory of the Creator. However, people cannot manifest this archetype because they imagine a Messiah who belongs exclusively to them. That's why tension is growing and will continue to grow until many of us will start seeing the Messiah in our hearts – as Spirit that is within, and for everyone. People are waiting for the Messiah to come to them, whereas they have to come to the Messiah in their heart and mind. We must do our own travelling to approach the centre of the whole experience of illumination Within.

A possible way of coming to unity lies in an understanding of the nature of any conflict from the standpoint of unity. This understanding is not always easy to come by because it is usually obscured with clichés and other prejudices that the powers from both sides of the conflict instil in the minds of their followers to manipulate them into obedience.

To escape this trap, we have to undertake a comprehensive

and unbiased research of the matter at hand. We have to ask and try to truthfully answer as many questions as possible. These questions should be asked from different angles of vision. The answers we receive should govern our actions.

This is not an easy task because, as *Agni Yoga* explains: "Circumstances may appear one way on the surface, but the true, deeper meaning may be entirely different. Sometimes the predestined is manifested in an unexpected guise. Thus one becomes used to the idea that justice has many faces."[7]

However, if we are determined to understand a conflict and eventually come to unity, we can achieve our goal by searching for the deeper meaning of the events. It helps when we expand our consciousness through the process of asking and answering questions and through sincere learning from our experiences, and from those of our predecessors throughout the history of humanity. Through our efforts we can achieve the consciousness of Wisdom, and, as a result, our mind can find the best ways for resolving the conflicts that obstruct the way towards peace. The more Light of real Truth we let into the conflict, the sooner we will come to its resolution.

Co-Measurement – the Path to the Future

Where is the Way to Balance?: Consciousness is primary in our understanding of creation. The Creator is creating worlds by the Fire of Its Consciousness and every co-creator is creating their world by the fire of their consciousness. It appears that we live in a world created by the power of our united consciousnesses.

Depending on the state of our consciousness there can be either a chaotic world, as a consequence of our war-like thoughts and misjudged values or an harmonious world where everyone is co-creating with others, in harmony according to their inner balance tuned to the balance of the entire Creation. How can we find this balance and be sure of it?

We are used to a balanced diet, balancing our accounts and other aspects of our material life which we can measure, count and compare. We usually have in mind personal or collective means of co-measuring and comparing the components of our balance. Nowadays we come face to face with the fact that any measurement of power, value and possibilities is connected with 'money'.

The Artificial Measure of Money: Money is the product of human intellect. It is a completely artificial, man-made instrument, reflecting human understanding of the Solid World. It can be compared to a school student's 'ruler' that provides measurement in a simple linear fashion. Just imagine what would become of our modern science and technology if a straight ruler remained 'the Ruler' of any measurement and no other instruments and measures were invented and applied to the complicated environments of human activity!

The instrument of money, which human intellect created initially only for the exchange of goods, is accepted and applied today as a universal measure. Of all the factors that could be taken into account in the process of decision making, economics assessed in terms of money is seen as the main factor. It affects and often dictates all decisions and actions of governments and other powerhouses on Earth today. Such an exceptional position for economics, only one of the many important aspects of human existence, leads to an imbalance in our relationships with each other, with nature, and with our own health and wellbeing. It is clouding our perception and leading to the misjudgement of values on a large scale, which denies us the ability of correctly evaluating our decisions.

So what can we do to direct our consciousness to create a more harmonious future for the world? We would like to pursue the best values, but we have to know which values are in harmony with the archetype reflecting the balance and beauty of highest manifestations, and which are obsolete and dragging us backwards. We have to know how to measure our values and how to co-measure them with the highest manifestation of Beauty. This leads us to discuss the Law of Co-measurement.

The Law of Co-measurement: In the perfect picture in the Memory of the Creator, everything is balanced and constantly self-balancing in conformity with the Creator's Consciousness. Everything there matches with perfect accuracy. When we see the reflection of this perfect match here in our Solid World, we often call it 'coincidence' or 'miracle'. When we receive the exact manifestation of the perfect balance in the Solid World, it is always a manifestation of Beauty. If we could only know how to manifest such an exact reflection with our actions! It would certainly help to solve our problems. According to *Agni Yoga*, this would require us to follow the cosmic Law of Co-Measurement.

Co-Measurement relates to our ability to evaluate the meaning and importance of every aspect of Life, which leads to right judgment in all our choices and deeds. As the Master says: "Co-measurement is like the pillar that supports the house." He gives us some hints of why we don't always produce Beauty and what may be not in conformity with correct co-measurement in our actions:

> *Now let us conclude the law of co-measurement. The builder must know how much load the pillars of the house can carry. From non-co-measurement results destruction, blasphemy, lie, treason, and many other ugly manifestations.*
>
> *Can a building endure wherein to a flea are attributed the qualities of a giant; wherein a poker is sought more than the Lord; wherein the whirlwind is being compared to a mosquito's flight?...*
>
> *...The best judgement will be founded upon beauty. It is ugly to say, "I shall put the giant into a little box", or, "The eagle soars like a hen".*

How often are the best apparatus destroyed by non-co-measurement, which, with a little attentiveness, is easy to avoid.[8]

The Master's words emphasize that it is our exaggerated attention to the petty issues of the lower ego that prevents us from correct co-measurement in our choices and actions. He assures us that "the smallest thoughts will be borne away by the whirlwind of right judgment. Good or bad, useful or harmful, these will be singled out, because where the big trees are, the shrubs are not seen."[9]

The Master advises us: "If we apply our efforts to change the hustle and bustle into a beautiful achievement, then the gnarled thorn bushes will be transformed at once into a tall grove. If we can rise in thought to the boundaries of the miraculous, then we shall not speak lengthily about a worn-out sole."[10] He warns us against overestimating our habitual values and calls us to make our spiritual values paramount over all worldly and mundane considerations. Spiritual achievement appears to be of the highest value.

This statement can be understood also in terms of the energy and frequency of energy bodies making the whole constitution of a human being (see "The Universe in the Human Body").

Connection between Energy and Vibration Frequency in Terms of Physics: As we now understand in the 21st century, energy is the source of the existence of every solid body. The waves or vibrations of this electromagnetic energy are filling every atom and cell of our bodies. The discovery made in 1900 by the physicist, Max Planck, on the value of the energy of a radiating body was later theoretically associated with the energy of the electromagnetic wave itself. It was discerned that the energy of a wave is directly proportional to its frequency.[11] And we can say that the energy that a body possesses is directly proportional to the frequency of its waves (vibrations).

The higher (closer to the Fiery World) our energy body the higher is also the frequency of its vibrations. The higher these frequencies the more energy (power) they carry. The higher the level that our thought can reach the more powerful it is and the quicker it can come to realization. This explains the Master's statements on levels of spiritual achievement in the terminology of modern physics.

Agni Yoga also states: "Each force which reveals a new possibility for the Common Good also manifests cosmically a channel for other revelations".[12] This statement opens the way for us to expect new possibilities to open for every one of us when we apply our higher frequency powers for the Common Good. Is this not a way towards the mastery of the direct manifestation of thought?

The Law of Co-measurement Stems from the Cosmic Laws of Equilibrium: *Agni Yoga* speaks about the cosmic Laws of Equilibrium which uphold harmony in all the three worlds: the Fiery, Fine (Subtle) and Solid ones. The cosmic laws of Equilibrium have multiple aspects and are multidimensional. We can find clues to their different aspects in the books of *Agni Yoga*, especially in part III of *The Fiery World*.[13]

The events in the cosmos are being attracted to designated dates, or regular cycles in the greater life of our Earth as a living organism:

> *The cosmic laws of Equilibrium are applicable to all dates on the planet. Spatial solutions are attracted to the designated dates and can be manifested in all the affirmations of life. Equilibrium is saturated with these spatial solutions. Therefore designated dates must be subtly calculated. One may study the map of the World in different epochs and perceive how the Cosmic Scales have manifested great Equilibrium.*[14]

According to *Agni Yoga*, all the cycles are formed in the spirals of Cosmic Ordinances. A new epoch is created when new energies are assembled and new beginnings are affirmed:

> *The Equilibrium of the World rests upon the foundation of Being. So powerfully is life affirmed when the higher manifestation is kept in the consciousness. Each lofty thought will be a pledge of the spirit's striving. And in an endless chain of action and thought can be expressed all new trends of evolution. Space affirms its tensions which conform to the actions and thoughts engendered on the Earth. This emphasizes how responsible is humanity for all its engenderings, for the Subtle World is thus held back in its development, just as is the whole chain of evolution on the Earth. Therefore thought about spirituality must enter into life, but as a true understanding of the Fundamentals of Existence. The Equilibrium of the World cannot be established without true understanding of the First Causes. Thus, each fiery word of the heart proceeding towards purification of the Teaching is a fiery stimulus which will give impetus to the shift of consciousness. Therefore, let us be affirmed in the consciousness upon the power of Equilibrium, as the stimulus of Existence, of the First Causes, and of Beauty.*[15]

It appears that the cosmic Law of Co-measurement, which stems from the Laws of Equilibrium and should be working throughout all the three Worlds, is not always upheld in our Solid World. This lack of harmony with the Cosmic Law allowed money to rule our thoughts and actions because we humans cannot see any purpose in life beyond the mere basic considerations of life in the Solid World.

To the majority of people the movement of shares on the stock market seems to be of more importance than the movements of stars! Market strategies based on the criterion of maximal profit define the ways in which societies are searching for their prosperity, most often at the expense of the well-being of people and Nature.

Moreover, the profits of weapon producers and weapon traders come from wars which result in the loss of human lives and destruction of livelihoods. Which, in its turn, produces a flow of refugees who have lost not only their livelihood and the positive input they used to create for their community, but also their family members and friendly environment, their jobs, carriers and use of their native language. They are left in a situation where they have to rely on the assistance of other countries and their people.

When using the criterion of maximal profit, we are creating contradictions, for example, with food production and distribution. Masses of people are starving whilst others are becoming more obese in many countries and masses of food goes to waste. Isn't this an inharmonious and simply crazy situation? Another contradiction: on the one hand, most profitable corporate farming and food production allows the production of more food. On the other hand, their profit-oriented ways of production are often destructive to nature and to human health. Also, they are detrimental to the survival of farmers and the old, healthier way of farming. No one can judge with conviction which way is right for humanity because the criterion of profit doesn't allow for it.

It therefore follows that we should change our ways of judgment and find new ways of co-measurement which would eliminate the inequity we see everywhere on Earth. Money is working only in, and for, the Solid World. The new criteria that we choose for the new epoch should reflect the realities of all the three worlds to which humanity belongs: the Fiery World, the Fine World and the Solid World. They should be real and connected with spiritual values such as Light, Beauty, Love and Wisdom. And they should allow a balanced relationship between humans and the rest of life.

The Measure of Human Value: When we start looking at the very beginning of everything, we see only the creating rays of Light. The Power of Light is the measure for everything that It creates. Finding its way through the three worlds of existence, the 'Silver Thread' – the animating Ray of the Creator – fills the heart 'chalice', the Fine World organ behind the human heart. It is in the heart chalice where Wisdom is born from the fusion of Spirit and Intellect. It is there where the trinity of Power, Love and Wisdom can be balanced.

Let us see what the Ascended Master, the alleged Teacher of both Helena Blavatsky and Helena Roerich, says about the future through *Agni Yoga*: "The Teaching of the Future Epoch will be re-union of the spirit and intellect."[16] H.P.Blavatsky also stated in *The Secret Doctrine* that humanity is on its way to balancing intellect with spirituality. We just have to remember that spirituality is not the same as religion. A person following a religious tradition may or may not penetrate into the depths of their spirit through religion, so their devotion to religion may or may not be called 'spirituality'. A non-religious person may also demonstrate, sometimes even not being conscious of it, the highest examples of spirituality.

A great change of understanding is sooner or later coming to us. The old measures and criteria which assisted in the development of our intellectual abilities and achievements cannot satisfy us anymore for the purpose of the "re-union of the spirit and intellect". They should be replaced in our consciousness and in our life. Human value should be measured not by the 'depth' of the 'purse', but by the capacity of the heart.

As we understand it, the capacity of the heart defines the integral energy of all the energy levels, or bodies, of a human being. It is reflected by the energy of every level, or body, which in the times of Blavatsky and Roerich was known as 'mental' or 'psychic' energy. *Agni Yoga* predicted that the measure of 'psychic energy' would become the basis for the evaluation of human qualities:

> *Not so long ago one could have regarded such discussions as moral abstractions; but when psychic energy is evaluated then human qualities will become scientific values. Is not the possibility fascinating – to be able to establish a scale of qualities on an experimental basis?*[17]

Realization of such an idea would benefit everyone. Our health, wellbeing and very happiness depend on our levels of energy. What is more, willingly or otherwise, we are sharing our energy with our surroundings. We can either improve or ruin the energy 'climate' for those with whom we communicate in a family, at work and in society. Knowing the 'scale of qualities' and having the means of measuring it would provide us with a tremendous possibility of self-evaluation and self-improvement, so that everyone of us would be able to consciously raise ourselves to the harmony of co-measurement.

When humanity comes to real understanding of the Law of Co-measurement, it will be able to evaluate people's thoughts and actions from the standpoint of the highest spiritual criteria. People will learn to apply the finest instrument of their inner spiritual 'tuning fork' connecting them to the Creator within, for the purposes of co-measurement.

If the 'capacity of the heart' measured in the terms of 'psychic energy' becomes the main measure for the evaluation of human qualities, how will human actions be co-measured? Whatever the system of measurement, the appreciation of actions towards self-improvement and self-help or work in the fields of teaching, healing, creative arts, service to the community, or the direct production of goods maintaining human life will rise compared to what we have today. All of them result in the rise of the energy levels of separate individuals, of a community of people and entire humanity. A person's stimulus for work, besides any outward appreciation, will ensue from the awareness of their input into the community and from the rise of personal energy attracted by the work of choice.

Co-measurement and Human Endeavours: Can we bring any change into the current situation where our technology and our economy are 'hooked' on money? In which direction should we move to get out of ultimate dependence on the monetary criteria of assessment? Are there any other mechanisms and measures, apart from monetary, which could serve as indicators for co-measurement in the processes of human activity? How can such mechanisms be applied?

In our book, *The Laws of Life*, in the chapter on 'convertors', we presented a model of a simple converter and the criteria for its evaluation. The criteria for the evaluation of the efficiency (usefulness) of the converter and of the potential harm from the waste that it produces were suggested in the 1970s by the Russian scientist, P.G.Kuznetsov.

A simple converter (see fig.8) is a process turning the potential possibilities, the input, N_0, into real possibilities, the output, N_1, where N_0 are material substances, energy and order or information put into the process, and N_1 is the resulting product. The process also produces some waste N_2.

The chosen criteria are the 'converter's efficiency' and the 'factor of contamination', where:

$k=N_1/N_0$ – converter's efficiency.

$q=N_2/N_1$ – the factor of contamination.

This means that the more output of product we receive from the same initial input, the higher the value of the criterion and the more effective the converter. The more waste is produced as a by-product of the process, the higher the factor of contamination. When the factor of contamination is higher, the value of the process is lower.

To lower the factor of contamination and to simultaneously

receive a higher output from the process, we can organize chains of processes (converters) recycling waste and turning it into a valuable product as is shown in fig.8. Potentially, we can almost nullify the final amount of unused and unusable waste and subsequently get converter efficiency of almost 1, or 100%.

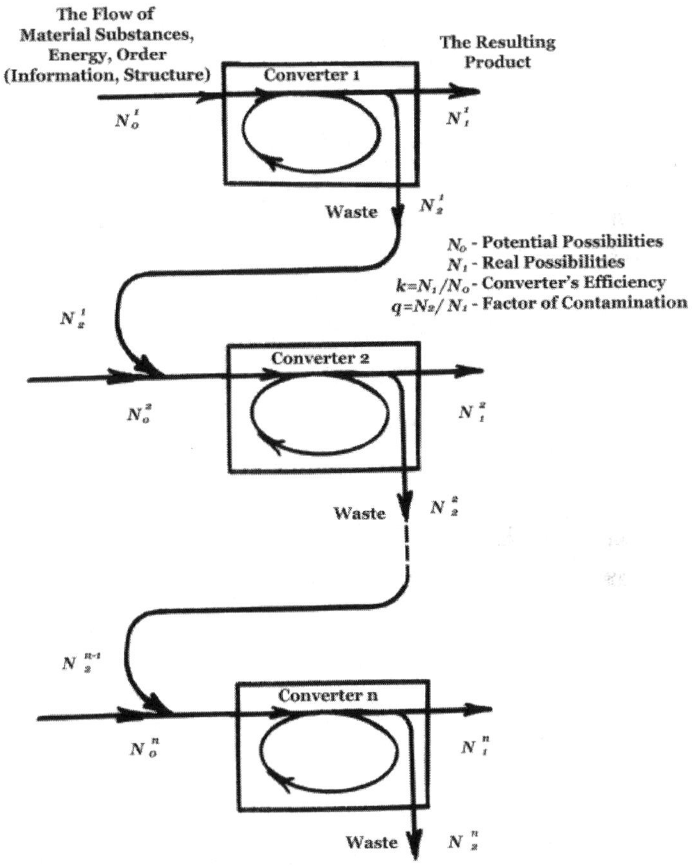

Fig.8. Chains of Converters

These criteria, if consistently applied, could lead to the creation of technologies based on higher frequency, clean and unlimited renewable sources, for example, solar energy. Of course, such criteria would lead us to concentrate on the highest frequency energy of thought – on vital new ideas.

By freeing itself from the fetish of money as the decisive criterion, humanity will release colossal resources of all types. It will be able then to direct these resources towards life sustaining and improving projects it cannot even dream of right now. Liberated from their focus on money, people will turn their attention to different kinds of creativity. And creativity means a constant search for balance and beauty.

Emulating the Universal Mechanisms: With the continuing growth of Earth's population, questions arise about so-called 'overpopulation' where the very ability of our planet to sustain its population is questioned.

We already have some solutions effectively turning renewable planet's resources into food products. There are working chains of recycling technologies, such as, for example, the 'Sundrop' farms technology of growing hydroponic fruits and vegetables in an Australian desert using solar energy and salty sea water.

If we are searching for ways of solving global problems like sustaining the world population, we have to make all technologies work in a co-measured way. Emulating universal mechanisms may help modern society to discover right solutions and new ways of development.

Looking at the universal laws and mechanisms of co-creation and evolution which are creating and sustaining life throughout the Manvantara of the manifest universe, we can see in them an exemplification of perfect balance and effectiveness. They manage to uninterruptedly provide life in the universe for a whole Manvantara! How do they do this?

When analysing the universal laws and their mechanisms, we come to their general attributes which are:

- universal connectedness,
- mutual dependence of cyclical processes represented by

cycles within cycles,
- inclusiveness, where every living essence is included and involved in reception and giving out of energy and information,
- balance of freedom of choice and obedience to cosmic laws for any living essence,
- constant motion and development of the whole system of manifested life and of its every part,
- unique ability of life to 'recycle' chaos and turn it into order and beauty (to create 'negative entropy').

Can we people somehow emulate natural laws and mechanisms and secure all the essentials of life and development?

Everything starts from consciousness, so to answer this question we have to start from making changes and expanding the abilities of our thinking. Especially important is the ability of turning chaos into order and beauty. While writing our book *The Laws of Life*, we came to the appreciation of a human being as a 'cosmic converter' that, "as a part of Life, has a purpose to transmute 'death' (understood as 'chaos', 'low energy vibrations', 'complete solidity' and 'ignorance') into the 'growth of Spirit' by uplifting consciousness and creating Beauty in every aspect of Life".[18]

Co-existence and collaboration of humans aspiring to such a high goal is impossible without trust. Even now, in our current society, any collaboration – in science, trade, family – can fully succeed only if it is built on trust. In the future society, which *Agni Yoga* calls 'The New Era Community', it becomes paramount. As *Agni Yoga* explains, such community already exists in the Subtle World as the 'Brotherhood' of Masters, 'a strong unity of trust'. The concepts on which this Brotherhood is built can be accepted as an example that future human society would emulate:

It may be that right now certain finer hearts are already dreaming about the creation of organizations where trust would be the cornerstone. It cannot be insisted that everything is bad, when the human eye sees only some of the details of the approaching epoch.

Upon the fragments of ancient symbols one may observe the vitality of the basic concepts. Just when, from the earthly point of view, everything has been transgressed, it may be that at the same time most beautiful concepts are already being born.[19]

Synthesis and Brotherhood

The Road to Unity: Nowadays, on our common level of perception, we think of ourselves as separate, completely independent personalities. In the boundaries of our secure Western society we make plans, reach our goals, enjoy our personal rewards and insure ourselves against ill fortune just in case. The very ability of achieving all of this individually, and seemingly without assistance, gives us the illusion of our separateness. Hence our individualism – the hallmark of Western Society since the 18th century.

Meanwhile, humanity as a whole is constantly undergoing different trials and tribulations, including wars, natural disasters and periods of financial crises, which periodically engross masses of people. Nevertheless the majority of humanity remains unable to understand that we are somehow united and dependent on the actions of each other.

We witness again and again human beings choosing to look at each other through gun-sights rather than amicably discussing their differences. Such inability to comprehend the wider picture makes them reflections of each other because they observe each other with similar self-centred thoughts. Most probably, even the colours and hues of their 'aura' are similar in situations of hatred leading to warfare.

There is a long and winding road upwards from this lowest point to spiritual evolution, gaining the ability of connecting causes and effects, advancing and adopting the ideas of cooperation and compassion. This is a road of experience and learning. On this road, the achieved mastery of synthesis of ideas and related facts and observations brings us to a completely new level of understanding. Further on, eventually, to the awareness of our Universal Unity.

Synthesis and Education: Throughout the ages, a person, while growing up, received their understanding of life either through the mysticism of ancient mysteries and religions or through modern 'hard core' science. The methodological difference between these two ways is that science studies the Solid World connections between phenomena, whilst real mysticism requires immersion into the vibrations of the subtler planes of reality.

If we accept and study the discoveries of science and are able to expand our consciousness to simultaneously see them in the light of a broader picture of Creation and Evolution throughout the Three Worlds of existence, then we can gain access to both methodologies and to the synthesis of their findings. In this book, we attempted to open up the common universal mechanisms in the light of such synthesis.

Synthesis is possible when we work not with the dogmas of either science or religion but with the physics and metaphysics of real mechanisms. In this case the mechanisms fit to each other and explain each other, so we can put together the whole 'puzzle' of the greater picture of consciousness creating everything in the universe. In this book, we attempted to open up the common universal mechanisms in the light of such synthesis.

In its book *Brotherhood, Agni Yoga* talks about the beauty of synthesis.[20] *Agni Yoga* explains that synthesizing means coming to an awareness of even the smallest particle of World Creation from a broader perspective of spiritual understanding. All of this broadens and adds dimensions to our observation. It is a road of experiences, lessons of karma and gaining of awareness through these experiences.

Through synthesis we come to Wholeness. We are on a Path to understanding that actually all of us have in our Essence this whole and perfect Thoughtform of the Creator. That every one of us cannot be whole without all the others. When we have this understanding we start to really respect all other people, who have come to their lives with their own special purpose, as truly our brothers and sisters. We realize the significance of all different branches of knowledge. We recognize and accept the concept of Universal Brotherhood.

Agni Yoga says:

> *Knowledge is the gateway to Brotherhood. Let us not be surprised that the establishment of Brotherhood begins with the synthesis of the sciences. Though each one master but one subject, nevertheless he will know how to render respect to the countless branches of knowledge. In such respect is born the understanding of Brotherhood.*[21]

Hence the utmost importance of education if we would ever hope to make the Earth a planet of harmony and abundance. For the people of the Earth to be able to gain abundance through harmony, the children, as the creators of the Earth's future, should be taught to synthesize knowledge.

Children have to get to know the mathematical language and the physics of the Creation of Life and its Existence in our world. They have to understand Evolution as it is going on in the Solid World, and the ways of development of all the Three Worlds. They have to learn to discover metaphysical and spiritual aspects of Life, at the same time perceiving arts and sciences as a united whole to understand the Unity of the beautiful picture of Existence and themselves as an integral part of Nature and the Universe.

The dimensions of their knowledge would then embrace all

the Worlds of Creation, all their laws and active mechanisms. They would become natural co-creators with the Creator, with all the ensuing power and responsibility for their actions, because simultaneously they would learn the art of achieving balance.

The Parable of the Rower

The Man came from afar. His memory refused to remember where he lived and what he did and saw there. It threw out the pictures of death and devastation. Actually, his memory rejected the past, the whole past universe in which he used to live.

The Man came to the shore where he could see only the boundless smooth surface of water ahead of him and a small boat with a pair of oars ashore. The Man pushed the boat off the shore and into the water, took the oars and started rowing. From then on, he got a new name – he became not just a Man, but the Rower.

He didn't know where he would finish; he just felt that he had to row forward, away from his past and into the future. Hour after hour, he continued rowing. Something gradually started to come of it. The rhythm of the oars and creaking of the rowlocks reminded him of a song, and little by little the Rower started singing. At the start he was just repeatedly and rhythmically murmuring some sounds, but eventually a melody came out of these sounds and a proper song appeared. The song was simple, with just a handful of words. However simple, the words carried hope.

We don't know how long the Rower was rowing and singing, he didn't know himself. Neither time nor any other signs existed in the azure expanse of the sea. Nevertheless, he observed the line of the horizon and gradually started imagining and singing about a beautiful island of his dreams. When he eventually saw the shoreline, at the beginning he thought that this was just a mirage. But the 'mirage' started growing and eventually turned into a real island.

Can we tell you that the island resembled the Rowers dreams? A sandy shore: then a jumble of greenery – grass, bushes and trees. A brook with clean, cool water, beautiful flowers and an abundance of edible berries in the bushes. The Rower knew that this was all he needed – his new universe. He could build the rest himself.

Later on some sailors from a ship noticed the previously unknown island, and it was eventually marked on the map. We don't know what name they gave to the island, but we call it 'the Rower's Island'.

A question remains unanswered – could the Rower create a real island – his universe – out of the rhythm and hard work of rowing, the song of his dream, and observation? Could he re-create his own universe and bring his dream into manifestation? Can we?

Conclusion

We look up into the night sky in wonder of the millions of stars, trying to fathom the secrets of the Universe. In this book, we have had a glimpse in the opposite direction – from the state of Eternity and Infinity towards our present condition in space and time. This reverse perspective offers us the possibility of acquiring a systemic understanding of the processes that manifest the Universe and our life.

We have tried to synthesize spirituality and science – and found that this method has huge potential for further research. It allows a scientist to use all the wealth of humanity's knowledge and all the possibilities of receiving fresh ideas by looking at the subject of research from a new perspective. It provides the opportunity of re-evaluating and appreciating the discoveries of sciences in any era.

On the other hand, the discoveries of modern science allow the researchers in metaphysics and spiritual philosophy to revise their ideas in the light of modern scientific knowledge.

Here we are, in this moment of time – in this universe, and in our shared humanity. In a sense, we are immensely rich: we have the whole universe as outside, as within us. We have an ocean of the purest, finest, and most powerful energy of Fire and Light at our disposal. We have consciousness by which we can reach the highest summits of awareness and creativity. We are unstoppable because nobody and nothing can prevent us from creating the most harmonious and beautiful world in us and around us – except ourselves!

The only obstacles that could stop or slow down our unceasing motion towards Light are some of our choices and decisions as Observers. What kind of a world do we imagine for ourselves, what goals are we seeking, and which intentions are we putting into our thoughts?

Whatever they are, we are learning. On the path of learning, we are crawling, then walking, then running, and only then, flying. We meet our teachers and become teachers ourselves. We have the power of rhythm to bring into empathy with us those whose vibration frequencies are close to ours. We can do it by sharing our energy and by our example. This is our chance to change the world for the better. Our power rises as soon as our thoughts are clear, and our energy vibration frequency becomes higher. The challenge lies before us – TRY!

Appendix 1

Agni Yoga. Review by Roza and Margarita Riaikkenen
Published in Theosophy Downunder, Newsletter of the Theosophical Society (Pasadena) Australasian Section,
No: 110 August 2013

Yet again we are turning the pages of the books of *Agni Yoga*. Like many times before, at the turning moments of our personal life or of the world around us, we are searching in these books for solace, encouragement and wisdom. And we are finding all we searched for and much more.

Agni Yoga, which is known also as the teaching of *Living Ethics*, is a series of books recorded by Helena Roerich, channelled from her spiritual Teacher. The books were dictated to Helena Roerich starting from 1920 and were recorded presumably until 1941. *Agni Yoga* operates with the finest fiery energy of the universe – Agni.

Helena Roerich, the wife and collaborator of renowned artist, Nicholas Roerich, was a remarkable representative of Russian mystics in the line to which Helena Blavatsky, the founder of the Theosophical Society, belonged. There are other similarities between the two Helenas besides their names. Both of them were Russian women who received some of their esoteric experience and knowledge in the Himalayas area, and both had the same Ascended Master as their spiritual Teacher.

Agni Yoga is a teaching of spiritual philosophy, a teaching of synthesis that embraces science, ethics, arts and culture, Oriental and Western, from the point of view of esoteric knowledge. It gives a multi-facet view of life, on Earth and in different places and dimensions of the universe and lights up many aspects of universal Truth.

The very goal of *Agni Yoga* is to show us the way of inner

transformation and evolution into the higher state of spirituality. You can just open any of the books and find advice on your spiritual path, on health and spiritual practice, on your possible psychological problem with explanation how it is connected with the current world condition, and much more.

Agni Yoga doesn't feed us with niceties, but we can find there keys to a lot of our questions. It is a rigorous teaching of selflessness, courage, compassion and sacrifice. And also of collaboration and co-creation. Of collective action and personal heroism, of scientific discoveries and self-discovery, of aspiration and achievement. And also of Joy from the participation in this inspirational work, spiritual joy, fiery joy: "Understand the beneficence of fiery joy. It expands the consciousness, and draws all the best accumulations toward itself, as a fiery magnet. The essence of man is transformed and his outworn nature is burned away" (*Supermundane IV*).

This is a teaching of esoteric science, with its strict laws and mechanisms, with the abundance of treasures of wisdom here and there. And the Heart (the Heart Chalice behind the physical heart), full of Love energy, is proclaimed as the source of everything a human being is able to achieve.

Agni Yoga gives us also some glimpses into the future of humanity, and it is of particular interest because the Master is speaking from the strata of existence where time is different than here in our Solid World. His position therefore allows him to see the future much more clearly than we could do, though we also have the potential to become conscious observers and co-creators of the future. However, "Only devotion and understanding of the wisdom of the Plan insure the reality of the pictures of the future" (*Leaves of Morya's Garden, Illumination*). Understanding of the Plan helps us to become aware of the wholeness and meaningfulness of our life with its past, present and future.

Appendix 2

The Fire of Kalagia. Review by Roza and Margarita Riaikkenen Published in Theosophy Downunder, Newsletter of the Theosophical Society (Pasadena) Australasian Section, No: 122 March 2017

The book *Kalagia* burst into the Russian spiritual life in 1993, with its exact definitions and mechanisms of the concepts that before haven't been so clearly explained either in the scientific or in the esoteric literature. What is the Source of life and what is Life itself, what are Time, Energy and what is the creative Fire within any living entity? Why should we search for this Fire? After reading *Kalagia,* we begin to better understand these concepts and also receive a fresh outlook on the classics of the esoteric literature of the past.

In *The Secret Doctrine* by H.P.Blavatsky (principal founder of the Theosophical Society in 1875), we can find the word 'Kalagni', where 'kal' originates from the Sanskrit 'tomorrow' and 'Agni' is the God of Fire in Vedas, the exemplification of the finest vibrations of the inner creating and consuming Fire.

Why 'Fire', and why 'tomorrow'? We can find some clues to this enigma when we remember that Helena Blavatsky's writings reflected messages which she received from the Masters of the Hierarchy of Light, from the World of high, fiery vibrations. Her duty was to pass them on to mankind, as was the duty of the later messengers who came in the 20th century, such as Nicholas and Helena Roerich, through the teachings of Agni Yoga. Before receiving and recording these teachings in the snowy Himalayas, Nicholas Roerich heard the call of the Hierarchy of Light: "Kalagia!" which he understood as "Come to Me!"

Should all of us come to the Hierarchy of Light? And when should we come? Tomorrow? How can we do this?

Half a century later, in the 1980s, a young Russian artist, Alexander Naumkin, who lived then in South Siberia, in the Altai region, also received a call from a Master of the Hierarchy. He had to climb high to the holy mountain of Belucha, build there a hut, and in this remote place write down the teachings of 'Kalagia'.

It is not surprising that the new teachings have been received in Russia. Russia can be likened to a high pressure furnace or crucible. It is difficult to live there, and a seeker for spiritual Truth may be exposed to very harsh conditions; but perhaps the crystals of true knowledge can be passed to the world only through the devoted hearts which are ready to endure tough experiences. So Alexander Naumkin had to go through the experiences of solitude and constant pain caused by direct contacts with the World of high, fiery vibrations.

The book *Kalagia* complements the books by H.P.Blavatsky and N. and H.Roerich, with exact esoteric scientific descriptions which were impossible to pass on through the earlier sources because at that time science hadn't yet opened the concepts which could allow people to understand these descriptions. *Kalagia* depicts the picture of the creation of the universe by the creating Spirit, with Fire and Light. It opens to us the mechanism of the Monad, a 'cell' of Life, as a 'reactor of evolution'.

This is a real reactor, filled with the primordial Fire. We all carry this spiritual Fire within, even if we don't know or care about this. We have covered this Fire, our spiritual Essence, with the veil of Maya, Illusion, and are going through our experiences in the solid material world as if it is the only reality. Clouded in this Illusion, we only see ourselves as mortal, 'one-lifetime' people, and all our efforts are directed to further augment our possessions in the material world, to consume and entertain ourselves in this world.

Kalagia shows us that in reality we are immortal, that "Life is the endless transmutation of death into the growth of Spirit, into the concentration of Its possibilities. This is an aspect of Matter's spirituality – aspiration to the future, and, in this way, the formation of the ingredients of Order from Chaos".

Kalagia predicts that 'tomorrow', when we come to the awareness of our immortal, inner, fiery essence, we will return into the Fire of our 'reactor', accomplishing our evolution. We will re-orient our consciousness from consumption to giving. We will voluntarily, with Love and Joy, sacrifice our illusory possessions for the possibility of our evolution.

Kalagia has never been translated into any language other than Russian. To do this, its knowledge should be perceived in its fiery essence, let through and passed on. We have made a humble effort of translating some of the main principles and mechanisms of *Kalagia* in our books *The Laws of Life* and *The Return of the Prodigal Son*. We tried to connect the wisdom of *Kalagia* with that which we received from the earlier sources, including *The Bhagavad Gita, The Holy Bible,* the writings by H.P.Blavatsky, N. and H.Roerich and other messengers, and the insights that we received from within when going through our own experiences of life.

Appendix 3

The Mysterious Story of X7. Exploring the Spiritual Nature of Matter. Review by Roza Riaikkenen
Published in Theosophy Downunder, Newsletter of the Theosophical Society (Pasadena) Australasian section,
No: 109 April 2013

What can be more valuable and compelling than a real spiritual experience, well documented and explained? We are usually treasuring the bits and pieces of such information when reading spiritual writings, and often guessing what kind of events led to it. And here we have a report in a form of a diary, spanning for almost 3.5 years: from the 28th of December 1953 to the 24th of May 1957. A report of unique spiritual experiences described in detail, understood and explained!

The contents of this book is presumed to be reported by a group of anonymous prisoners of conscience incarcerated in a salt mine in Siberia – one of many of the Stalin era GULAG prisons. Out of the despair of their hopeless situation, the X7 (as they call themselves) came to an earnest prayer, asking why they were so entombed, was there any purpose to their ordeal or were they just God forgotten unlucky ones. Not at once, but gradually, they saw the Light that eventually liberated and transformed them into explorers of the spiritual nature of Matter. They also felt themselves being under the protection and guidance of the Presence of a grand Master of Light, who explained to them their purpose and duty in their current incarnation.

It sounds, at the start of reading, as if it "cannot be", but the further you go, the more compelling the text becomes. When showing the phenomena of Light radiations that became simultaneously their object of experiment and its tool, the authors recognize these radiations, or Light Rays, as the very source of everything, and explain some mechanisms of how they work. X7 demonstrate knowledge of mechanisms and processes that exoteric science is still unaware of, even now, though there are some hints from esoteric sources that this is so. And the X7 not only theorize, but practically live this knowledge in their supposed darkness inside the Earth.

What is more, they experimented with the Earth itself. Applying their new gained abilities, they sent radiations to melt the Earth's substance into light and then create forms of matter out of this light. These rays also healed and nurtured them, and they were not much in need of coarse, chemical processes of food consumption to turn into energy for their sustenance.

What was new and very valuable for us – they researched and explained in detail the colour-sound properties of the radiations in connection with their potential for those who are able to perceive them. The whole spectrum of colour and tone (which results from energy vibrations of different frequencies) can be divided depending on their functions, each of which is allotted a colour and tone. For example, the colour blue has a sustaining quality – the major quality through which creation is held in form.

Every living being on Earth manifests certain qualities of radiations which formed it. For example, a tree demonstrates its 'treeness' as something strong yet growing and changing. We people, as living beings, are also constantly absorbing radiation of different colours and tones, and the very quality of our life depends on their combination which we are receiving and giving out. As the X7 state, "the Ray of Love is the greatest because it has the essence of all things within it, and from it all things are made."

A lot of wisdom is coming from this little book, which we would recommend to every Seeker for universal Truth. A lot of findings for everyone, and amazement – how much exertion is necessary to come to such understanding and abilities!

And at last, how this supreme knowledge could be transmitted to the world from the Siberian mine? It appears that the prisoners transmitted their thoughts, and a sensitive medium from the "Network of Light" group at Findhorn, Scotland, UK, received their contents telepathically. After a couple of decades of indecisiveness by the people who had access to the text, it has eventually been published by Findhorn Press in 1979 and republished in 1996 and 2010, and now we have gained access to the extraordinary findings of the group of prisoners who called themselves simply X7.

Notes

Introduction

1. Bohm, David, *Science, Spirituality, and the Present World Crisis* (presented at the 12th International Transpersonal Association Conference, June 20-25th, 1992).
2. Bohm, David, *Wholeness and the Implicate Order,* 2002, London, Routledge.
3. Bohm, David, *The Holographic Paradigm and other paradoxes,* 1993, Boston, Shambhala Publications.
4. Pribram, K.H., *Brain and Perception: Holonomy and Structure in Figural Processing,* 1991, Lawrence Erlbaum Associates.
5. Witten, Edward, *Duality, Spacetime and Quantum Mechanics,* KITP public lectures, 1998, available at http://online.kitp.ucsb.edu/plecture/witten/ as on the 1/8/2016 as on the 13/3/2017.
6. Rosenblum, Bruce, and Kuttner, Fred, *Quantum Enigma,* Sec. Ed., 2011, Oxford University Press, 130.
7. Ibid., 148.
8. Ibid., 220.

Chapter 1. *The Process of Manifestation – Creation*

1. *Agni Yoga, Infinity,* 73, 1997, New York, Agni Yoga Society. (*Agni Yoga,* which is known also as the teaching of *Living Ethics,* is a series of books recorded by Helena Roerich, channelled from her spiritual Teacher. The books were dictated to Helena Roerich starting from 1920 and were recorded presumably until 1941. Agni Yoga books have been translated into English under the author's supervision, and the translations are available in print, see bibliography, and online at the website of Agni Yoga Society http://agniyoga.org/index.php.)
2. Blavatsky, H.P., *The Secret Doctrine,* v.1, 17, 1977, Pasadena,

Theosophical University Press. (Strictly speaking, the word 'Manvantara' does not exactly mean the Creation of the Universe. According to *The Secret Doctrine,* this is a period between the two consecutive 'Manus' – the entities which appear first at the beginning of any manifestation. There are many types of Manvantaras: universal, of the solar system, planetary, and of man. In this book, unless said otherwise, we mean the universal Manvantara.)

3. Ibid., v1, 44.
4. Naumkin, Alexander, *Kalagia,* 1.10. 3 and 1.10.9, 1998, Kaunas, Gyvata. Published in Russian. (*Kalagia* is the book of "The Teaching of Evolution of a Human Being and Universe accepted on the Mahasatyana's Ray from the Teacher of Humanity Jesus Christ – Maitreya by the chela A.P.Naumkin in the mountains of Altai", in 1992-1993. *Kalagia,* as the continuation of previous genuine esoteric teachings, describes the principles and precise mechanisms of human power over time and evolution from an ordinary terrestrial man into a cosmic man. *Kalagia* is written in Russian. While translating from Russian to English, we have attempted to preserve the original spirit of the words. As to the word – "Kalagia": the esoteric writer and artist Nicholas Roerich heard the call "Kalagia!" while travelling in Tibet and the Himalayas and understood it as "Come to Me!" *Kalagia* explains this call as "Come Here!" and "Victory over the Power of Time!" *The Secret Doctrine* by H.P.Blavatsky mentions "the flame of time, Kalagni". As we can understand from these books and also from the books of *Agni Yoga* by Helena Roerich, advanced spiritual entities make their call and encourage us forward on the Path of our spiritual Evolution to the unity in the Fire of our inner Spirit, where we can gain our victory over Time.)
5. Ibid., 1.10.3. (Why is the density of Time measured by galvanometer? Galvanometer is a device for measuring

electric current using its electromagnetism; and the electromagnetic wave of Light is the source of Matter whereas Time is the inner space of Matter.)

6. Ibid., 1.22.3.
7. Ibid., 1.27.1.
8. Blavatsky, H.P., *Isis Unveiled,* v.2, 39, 1988, Pasadena, Theosophical University Press.
9. Blavatsky, H.P. *The Secret Doctrine,* v.1, 278, 1977, Pasadena, Theosophical University Press.
10. Fortune, Dion, *The Mystical Qabalah, 2000, Red Wheel/Weiser.*
11. Blavatsky, H.P. *The Secret Doctrine,* v.1, 206, 1977, Pasadena, Theosophical University Press.
12. Naumkin, Alexander, *Kalagia,* 1.15.2, 1998, Kaunas, Gyvata. Published in Russian.
13. Blavatsky, H.P. *The Secret Doctrine,* v.2, 112, 1977, Pasadena, Theosophical University Press.
14. *Agni Yoga, Infinity II,* 223, 1997, New York, Agni Yoga Society.
15. Schneider, Michael. S., *A Beginner's Guide to Constructing the Universe,* 72, 1995, Harper. (We can also find descriptions of this process of 'weaving' in the *Ancient Egyptian Mythology* by Jo Forty and in *Esoteric Egypt* by J.S.Gordon. All in all it appears that Neith was the goddess of weaving, war and wisdom).
16. Naumkin, Alexander, *Kalagia,* appendix, 1998, Kaunas, Gyvata. Published in Russian.
17. Ibid., 1.10.32.
18. Ibid., 2.1.4. (We also explain the mechanism of Time in our book, *The Laws of Life,* 22-26, and in the lecture on Time available at http://www.theosophydownunder.org/library/theosophical-lectures/the-nature-of-time-by-roza-and-margarita-riaikkenen/ as on the 13/3/2017.)
19. Riaikkenen, Roza & Margarita, *The Laws of Life,* 13-52, 2004, Victoria, Canada, Trafford Publishing.
20. Rosenblum, Bruce and Kuttner, Fred, *Quantum Enigma, 4,*

2011, Oxford University Press.

21. Blavatsky, H.P. *The Secret Doctrine,* v.1, 179, 1977, Pasadena, Theosophical University Press.
22. Naumkin, Alexander, *Kalagia,* 1.16.3, 1998, Kaunas, Gyvata. Published in Russian.
23. Ibid., 1.16.2.
24. Ibid., 1.14.9.
25. Riaikkenen, Roza & Margarita, *The Laws of Life,* 14-15, fig.2, 2004, Victoria, Canada, Trafford Publishing.
26. Ibid., 212 – 215.
27. Calleman, Carl Johan, *The Mayan Calendar and the Transformation of Consciousness,* 120, 2004, Rochester, Vermont, Bear & Company. (The ancient peoples of America, most notably the Mayans, had a complex cyclic-spiral Sacred Calendar, 'tzolkin', providing them with a "unifying pattern for all creation". The cyclic-spiral movement is actually reflected in our everyday vision of Time progression based on our knowledge of the rotation of Earth around its axis. So in our clocks every 12 hours we return to the same point in time but a half-day forward, or every 24 hours – a day forward. We also use our knowledge of the rotation of Earth around the Sun in our calendars – after every 365 or 366 days we return to the same date but a year forward).
28. Rooke, Andrew, T*he Sun: Powerhouse and Father/Mother/Elder Brother, Sunrise,* August/September 2002.
29. Naumkin, Alexander, *Kalagia,* 1.6.1, 1998, Kaunas Gyvata. Published in Russian.
30. *Indigenous Traditions – Australia. Aboriginal and Torres Strait Islanders,* Library – Religion, ABC 9 July 2016, available at http://www.abc.net.au/religion/stories/s790117.htm as on the 18/3/2017.
31. Lawlor, Robert, *Voices of the First Day,* 36-37, 1991, Rochester, Vermont, Inner Traditions.
32. Ibid., 115.

33. Ibid., 267.
34. Blavatsky, H.P. *The Secret Doctrine,* v.2, 87, 1977, Pasadena, Theosophical University Press.
35. *Agni Yoga, Infinity II,* 212, 1997, New York, Agni Yoga Society.
36. Ibid., *Supermundane,* 97.

Chapter 2. *Manifestation of Nature*:

1. Gordon, J.S., *Esoteric Egypt,* 113, 2015, Rochester, Vermont, Bear & Company.
2. Schneider, *Michael. S. A Beginner's Guide to Constructing the Universe,* 115-118, 1995, Harper.
3. Ibid., 117.
4. Ibid., 119-128.
5. Naumkin, Alexander, *Kalagia,* 1.15.14, 1998, Kaunas, Gyvata. Published in Russia, excerpts translated by R. and M. Riaikkenen.
6. Schneider, Michael. S., *A Beginner's Guide to Constructing the Universe,* 234, 1995, Harper.
7. Lawlor, Robert, *Voices of the First Day,* 114-115, 1991, Rochester, Vermont, Inner Traditions.
8. *The Mysterious Story of X7. Exploring the Spiritual Nature of Matter,* 3-83, 2010, Scotland, UK, Findhorn Press. (*The Mysterious Story of X7* is a report of life experiences of a group of prisoners trapped in a salt mine in Siberia, Russia, recorded from December 1953 up to May 1957. Their story has been telepathically transferred to a group of sensitives called "Network of Light" in Findhorn, England, and eventually published in 1979 by the Findhorn Foundation; republished in 2010.)
9. Ibid., 13. (Russian composer, Alexander Skriabin, showed us an example of colour-tone creativity in his symphonic masterpiece Prometheus: The Poem of Light, op.60 (1910), composed in music and light, where the part for the colour organ added an extra dimension of light to this already

transcendent masterpiece of sound.)

10. *Agni Yoga, Infinity I, 73, 1997, New York, Agni Yoga Society.*
11. Naumkin, Alexander, *Synergetics, the Introduction to Kalagia*, 25, 1998, Kaunas, Gyvata. Published in Russian.
12. Living Root Bridges of Charrapunji, India, available at https://www.youtube.com/watch?v=mmmlgKr0TzM as on the 25/7/2016.
13. Riaikkenen, Roza & Margarita, *The Laws of Life*, 33, 2004, Victoria, Canada, Trafford Publishing.
14. Ibid., 134.
15. Ibid., 138.
16. Ibid., 131.
17. Lawlor, Robert, *Voices of the First Day*, 236, 1991, Rochester, Vermont, Inner Traditions.

Chapter 3. *Manifestation of the Human Body*:

1. Naumkin, Alexander, *Kalagia*, 1.15.6, 1998, Kaunas, Gyvata. Published in Russian.
2. Ibid., 1.15.4.
3. Cranston, Sylvia, *The Extraordinary Life and Influence of Helena Blavatsky Founder of the Modern Theosophical movement*, 209, 1993, New York, Putnam Book.
4. *The Mysterious Story of X7. Exploring the Spiritual Nature of Matter*, 20-24, 2010, Scotland, UK, Findhorn Press.
5. *Saint Germain on Alchemy: Formulas for Self-Transformation*, recorded by Elizabeth Clare Prophet and Mark L.Prophet, 1988, Livingston, Summit University Press.
6. Riaikkenen, Roza & Margarita, *The Laws of Life*, 257-261, 2004, Victoria, Canada, Trafford Publishing.
7. Naumkin, Alexander, *Kalagia*, 1.15.5, 1998, Kaunas, Gyvata. Published in Russian.
8. *The Mysterious Story of X7. Exploring the Spiritual Nature of Matter*, 22, 2010, Scotland, UK, Findhorn Press.
9. Naumkin, Alexander, *Kalagia*, 1.15.13, 1998, Kaunas, Gyvata.

Published in Russian.
10. Ibid., 1.15.16 and 1.18.5.
11. Schneider, *Michael S., A Beginner's Guide to Constructing the Universe,* 241, 1995, Harper.
12. Naumkin, Alexander, *Kalagia,* 1.15.15, 1998, Kaunas, Gyvata. Published in Russian.
13. Ibid., 1.15.14.
14. Lommel, van, Pim, *Consciousness beyond Life,* 174, 2001, Harper.
15. Naumkin, Alexander, *Kalagia,* 2.3.1, 1998, Kaunas, Gyvata. Published in Russian.
16. Ibid., 2.6.1.
17. Ibid., 1.3.4.
18. Blavatsky, H.P. *The Secret Doctrine,* v.2, 43, 1977, Pasadena, Theosophical University Press.
19. Gariaev, P.P., *Wave Based Genome,* 1994, Moscow, Ed.ObshestyPol'za. Published in Russian.
20. Gariaev, P.P., *Wave Genetic Code,* 1997, Moscow, Ed. Izdatcenr. Published in Russian.
21. Gariaev P.P., Chudin V.I., Komissarov G.G., Berezin A.A., Vasiliev A.A., *Holographic Associative Memory of Biological Systems.* Proceedings SPIE – The International Society for Optical Engineering, Optical Memory and Neural Networks, 1621, 280-291, USA, 1991. (Gariaev also speaks of the fundamental property of holography in cells – the quantum non-locality of the genetic information – which allows super-fast direct communication and coordination between cells. Hence his idea of a DNA-wave bio-computer mimicking the workings of a living organism.)
22. Gariaev P.P., Junin A.M., *Phantom Leaf Effect. Myth or reality?* Energy 10, 46-52, 1989. Published in Russian.
23. Trafton, Anne, *A New Dimension for Genome Studies,* MIT News Office, October 9, 2009.
24. Blavatsky H.P. *The Secret Doctrine,* v.2, 299, 1977, Pasadena,

Theosophical University Press.

25. Ibid., 296, footnote. (The pineal gland is a small endocrine gland that lies in-between the two halves of the brain on the back portion of its third cerebral ventricle. It is known that the pineal gland regulates our circadian rhythms of sleeping and waking conditions. It synthesizes and secretes melatonin, a hormone that provides the body with information about environmental lighting. However, spiritual sources allot to the pineal gland functions of greater significance. There are indications that the pineal gland enhances human higher faculties of mind and serves as a connecting tissue for brain's communication with the higher spheres of existence. As *The Secret Doctrine* says, it is also the acting organ of the so-called 'third eye' – a sensor for seeing in the Subtle World.)
26. Luke, Jennifer, *Fluoride Deposition in the Aged Human Pineal Gland,* 2001, School of Biological Sciences, University of Surrey, Guildford, UK, Department of Obstetrics and Gynaecology, The Royal London Hospital available at http://www.icnr.com/articles/fluoride-deposition.html as on the 30/7/2016. (Unfortunately, we sometimes unwittingly harm that what we are most in need of for our future wellbeing. This scientific research shows that fluoride which became a mandatory additive to tap water in many countries around the world accumulates in the pineal gland. We don't know as yet how it affects the functioning of this important organ of the brain, so more research is necessary.)
27. Naumkin, Alexander, *Kalagia,* 1.5.4, 1998, Kaunas, Gyvata. Published in Russian.
28. Planck, Max, *The Observer*, 25 January 1931.
29. Lommel, van, Pim, *Consciousness beyond Life*, 2001, Harper.
30. Pribram, K. H., *What is Mind that the Brain May Order It?, 1997,* Proceedings of Symposium in Applied Mathematics,v.52, 311.
31. Pribram K.H., *Brain and Perception; Holonomy and Structure*

in Figureal Processing, 73, 1991, Hillsdale, N.J. Lawrence Erlbaum Associates.

32. Pribram, K. H., *What is Mind that the Brain May Order It?, 1997,* Proceedings of Symposium in Applied Mathematics,v.52, 320.
33. Doidge, Norman, *The Brain That Changes Itself,* 188, 2008, Melbourne, Scribe.
34. Doidge, Norman, *The Brain's Way of Healing,* 169, 2015, New York, Viking.
35. Ibid., 107.
36. Ibid., 109.
37. Ibid., 24.

Chapter 4. Development of Life in the Universe and in Humans:

1. Purucker, G. de, *Occult Glossary,* 98-99, 1972. Pasadena, Theosophical University Press.
2. Theosophy calls human archetypes 'Manu' (the word 'Manvantara', or 'Manu-vantara', originates from 'Manu' and is a period between two 'Manus'. 'Manu', in its turn, originates from 'man' – 'to think' in Sanskrit, according to H.Blavatsky's *Theosophical Glossary*. Another glossary,The Encyclopedic *Theosophical Glossary,* available at http://www.theosociety.org/pasadena/etgloss/man-mar.htm as on the 13/3/2017, states that the 'Manu' "... not only give birth to humanity but to animals and all other creatures").
3. Blavatsky, H.P. *The Secret Doctrine,* v.2, 598, 1977, Pasadena, Theosophical University Press.
4. Riaikkenen, Roza & Margarita, *The Laws of Life,* 37, fig.5, 2004, Victoria, Canada, Trafford Publishing.
5. Purucker, G. de, *Occult Glossary,* 138, 1972, Pasadena, Theosophical University Press.
6. Blavatsky, H.P., *Isis Unveiled,* v.1, 152, 1988, Pasadena, Theosophical University Press.

7. Blavatsky H.P., *The Secret Doctrine*, v.2, 300, 1977, Pasadena, Theosophical University Press.
8. *Agni Yoga,Supermundane,* 97, 1997, New York, Agni Yoga Society.
9. Ibid, *Supermundane,* 97.
10. Riaikkenen, Roza & Margarita, *The Laws of Life*, 20, 2004, Victoria, Canada, Trafford Publishing.
11. Naumkin, Alexander, *Kalagia,* appendix, 1998, Kaunas, Gyvata. Published in Russian.
12. Riaikkenen, Roza & Margarita, *The Laws of Life*, 67, 2004, Victoria, Canada, Trafford Publishing.
13. Naumkin, Alexander, *Kalagia,* 2.4.3, 1998, Kaunas, Gyvata. Published in Russian.
14. Ibid., 1.27.1.
15. Riaikkenen, Roza & Margarita, *The Laws of Life*, 88-113, 2004, Victoria, Canada, Trafford Publishing.
16. *Agni Yoga, Leaves of Morya's Garden, vol.2 Illumination,* 225, 1997, New York, Agni Yoga Society.
17. Ibid., *Infinity I,* 70.
18. Rooke, Andrew, *The Presence of Light, Theosophy Downunder* Newsletter, March 2016, http://theosophydownunder.org/australiantsnewslettermarch2016.pdf.
19. Naumkin, Alexander, *Kalagia,* appendix, 1998, Kaunas, Gyvata. Published in Russian.
20. Ibid.,1.31.2.
21. Ibid.,1.31.12.
22. Ibid.,1.13.12.
23. *Agni Yoga, Supermundane 1,* 213, 1997, New York, Agni Yoga Society.
24. *The Mysterious Story of X7. Exploring the Spiritual Nature of Matter,* 2010., Scotland, UK, Findhorn Press.
25. Rooke, Andrew, *The Sun: Powerhouse and Father/Mother/Elder Brother, Sunrise,* August/September 2002.
26. *The Mysterious Story of X7. Exploring the Spiritual Nature of*

Matter, 23, 2010, Scotland, UK, Findhorn Press.
27. Ibid., 63.

Chapter 5. The Ways of Shaping the Future:

1. Lommel, van, Pim, *Consciousness beyond Life,* 30-31, 2001, Harper.
2. Riaikkenen, Roza & Margarita, *The Laws of Life,* 99 and 314, 2004, Victoria, Canada, Trafford Publishing.
3. *Saint Germain on Alchemy: Formulas for Self-Transformation,* recorded by Elizabeth Clare Prophet and Mark L.Prophet, 277, 1988, Livingston, Summit University Press.
4. Purucker, G. de, *Fountain Source of Occultism,* 1974, Pasadena, Theosophical University Press.
5. St. Germain, *Earth's Birth Changes* (St Germain through Azena), 261, 1993, Cairns, Triad Publishers PTY LTD.
6. *The Mysterious Story of X7. Exploring the Spiritual Nature of Matter,* 32, 2010, Scotland, UK, Findhorn Press.
7. *Agni Yoga, Supermundane,* 290, 1997, New York, Agni Yoga Society.
8. Ibid., *Leaves of Morya's Garden, vol.2, Illumination,* 208.
9. Ibid., *Leaves of Morya's Garden, vol.2, Illumination,* 316.
10. Ibid., *Leaves of Morya's Garden, vol.2, Illumination,* 316.
11. The Planck relation describes energy in terms of vibration frequency as E=hv, where 'E' is energy in Joules (J), 'v' is frequency in hertz (hz which is equal 1/s), and 'h' is the so-called 'Planck's constant' h=6.626x10^{-34} (Jxs).
12. *Agni Yoga, The Fiery World III,* 238, 1997, New York, Agni Yoga Society.
13. Ibid., *The Fiery World III,* 230-240.
14. Ibid., *The Fiery World III,* 231.
15. Ibid., *The Fiery World III,* 240.
16. Ibid., *Leaves of Morya's Garden,* vol.2 *Illumination,* 220.
17. Ibid., *Brotherhood,* 51.
18. Riaikkenen, Roza & Margarita, *The Laws of Life,* 83-84, 2004,

Victoria, Canada, Trafford Publishing.

19. *Agni Yoga, Brotherhood*, 5, 1997, New York, Agni Yoga Society.
20. Ibid., *Brotherhood*, 427.
21. Ibid., *Brotherhood*, 428.

Bibliography

Bentov, Itzhak, 1988, *Stalking the Wild Pendulum. On the Mechanics of Consciousness*. Rochester, Vermont, Destiny Books.

Blavatsky, H.P., 1977, *The Secret Doctrine, v.1&2*. Pasadena, Theosophical University Press.

Blavatsky, H.P., 1988, *Isis Unveiled, v.1&2*. Pasadena, Theosophical University Press.

Blavatsky, H.P., 1892, The *Theosophical Glossary*. London, The Theosophical Publishing Society.

Blavatsky, H.P., 1992, The *Voice of the Silence*. Pasadena, Theosophical University Press.

Bell, J.S., and Aspect, A., 2004, *Speakable and Unspeakable in Quantum Mechanics*, 2nd ed. London, Cambridge University Press.

Bohm, David, 2002, *Wholeness and the Implicate Order*. London, Routledge.

Bohm, David, 1993, *The Holographic Paradigm and other Paradoxes*, ed. by Ken Wilber, Boston, Shambhala Publications.

Calleman, Carl Johan, 2004, *The Mayan Calendar and the Transformation of Consciousness*, Bear & Company, Rochester, Vermont.

Christianson, Gale E., 1984, *In the Presence of the Creator: Isaac Newton & His Times*. New York, The Free Press, Macmillan.

Cladesdale, Ruth, 2009, *Secret Wisdom*. London, Arcturus.

Cranston, Sylvia, 1993, *The Extraordinary Life & Influence of Helena Blavatsky, Founder of the Modern Theosophical Movement*. Putnam Books, New York.

Dante, Alighiery, 2013, *The Divine Comedy*, translated by Clive James. WW Norton & Co.

Davies, Paul, *1995, About Time*. Touchstone.

Doidge, Norman, 2008, The *Brain that Changes Itself*. Melbourne, Scribe.

Doidge, Norman, 2015, *The Brain's Way of Healing*. New York, Viking.

Einstein, Albert, 2001, *Relativity*. London, Routledge.

Finkel, Michael, March 2014, *Star Eater*. National Geographic Magazine.

Fordham, Frieda, *1991, An Introduction to Jung's Psychology*. London, Penguin Books.

Fortune, Dion, 2000,*The Mystical Qabalah*. Red Wheel/Weiser.

Forty, Jo, 1996, *Ancient Egyptian Mythology*. London, Promotional Repring Company LTD.

Freke, Timoty & Gandy, Peter, *The Hermetica*, 1998, London, Piatkus.

Gariaev P.P., Junin A.M., 1989, *Phantom Leaf Effect. Myth or Reality?* Energy 10, 46-52, (Russian).

Gariaev P.P., 1994, *Wave Based Genome*. Moscow, Ed. Obshesty Pol'za, (Russian).

Gariaev P.P., 1997, *Wave Genetic Code.* Moscow, Ed. Izdatcentr, (Russian).

Gariaev P.P., Birshtein B.I., Jarochenko A.M., Marcer P.J., Tertishny G.G., Leonova E.A., Kaempf U., 2001,*The DNA-wave Biocomputer*. "CASYS" – International Journal of Computing Anticipatory Systems (ed. D.M.Dubois), 10, 209-310, Liege, Belgium.

Gariaev P.P., Chudin V.I., Komissarov G.G., Berezin A.A., Vasiliev A.A., 1991,*Holographic Associative Memory of Biological Systems*. Proceedings SPIE – The International Society for Optical Engineering, Optical Memory and Neural networks, 1621, 280-291, USA.

Gordon, J.S., 2015, *Esoteric Egypt*. Rochester, Vermont, Bear & Company.

Goswami, Amit, 1995, *The Self-Aware Universe*. Tarcher Perigree.

Grof, S., 1975, *Realms of the Human Unconscious: Observations from LSD Research*. N.Y.: Viking Press.

Grof, S., 1985, *Beyond the Brain: Birth, Death and Transcendence in*

Psychotherapy. State University of New York Press.

Hawking, Stephen, 2011, *A Brief History of Time (From the Big Bang to Black Holes)*. Bantam Books.

Hegel, G.W.F., 1977, *Phenomenology of Spirit*. Oxford, Clarendon Press.

Holy Bible: John 1:1.

Kuznetsov P.G., 2014, *Selected Works*. Dubna, (Russian).

Lawlor, Robert, 1991, *Voices of the First Day*. Inner Traditions, Rochester, Vermont.

Lao Tzu, 1993, *Tao Te Ching,* translated by Stephen Adiss & Stanley Lombardo. Indianapolis/Cambridge: Hackett Publishing Company.

Lommel, van, Pim, 2001, *Consciousness Beyond Life*. Harper.

Long, James A., 1999, *Thoughts and Their Influence, Sunrise,* v.48, N3.

Jung, C.G., 2009, *The Red Book*. New York, London, W.W. Norton & Co.

Luke, Jennifer, *Fluoride Deposition in the Aged Human Pineal Gland,* 2001, School of Biological Sciences, University of Surrey, Guildford, UK, Department of Obstetrics and Gynaecology, The Royal London Hospital available at http://www.icnr.com/articles/fluoride-deposition.html as on the 30/7/2016.

Marcus Aurelius, 1964, *Meditations,* translated by Maxwell Staniforth. Harmondsworth, Penguin Books.

McTaggart, Lynne, 2008, *The Field,* Harper.

Naumkin, A., 1998, *Kalagia* (Russian). Kaunas, Gyvata.

Naumkin, A., 1998, *Synergetic* (Russian). Kaunas, Gyvata.

Oderberg, I.M., October/November 1986, *The Other Newton, Sunrise.*

Patanjali, 1980, *The Yoga-Sutras of Patanjali,* English translation with an introduction by M.N. Dvivedi. Delhi: Satguru Publications.

Phillips, Stephen M., 1996, *Anima*. Adya, Madras, T.P.H.

Planck, Max, *The Observer,* 25 January 1931.

Planck, Max, 2010, *Treatise on Thermodynamics*. Dover Books on Physics.

Pribram K.H., 1991, *Brain and Perception: Holonomy and Structure in Figural Processing*. Lawrence Erlbaum Associates.

Pribram K.H., 2004, *Brain and Mathematics: Brain and Being*. At the Boundary between Science, Philosophy, Language and Art. John Benjamin's Publishing Company.

Pribram K.H. , 1997, *What is Mind that the Brain may Order It?* Proceedings of Symposium in Applied Mathematics,v.52.

Pullen, Walter D., 2011, *Evolution of the Spirit: Our Journey through the Universe*. Ozark Mountain Publishing Inc.

Purucker, G. de, 1974, *Fountain Source of Occultism*. Pasadena: Theosophical University Press.

Purucker, G. de, 1972, *Occult Glossary*. Pasadena: Theosophical University Press.

Riaikkenen R&M, 2004, *The Laws of Life/The Return of the Prodigal Son*. Victoria, Canada, Trafford Publishing.

Roerich, Helena, 1929 (Republished 1997), *Agni Yoga*. New York, Agni Yoga Society.

Roerich, Nicolas, 1939, *Banner of Peace* available at www.roerich.ee/articles/eng banner of peace.htm as on the 30/7/2016.

Roerich, Nicholas, 1978, *Shambhala*. N.Y.: Nicholas Roerich Museum.

Rooke, Andrew, August/September 2002, *The Sun: Powerhouse and Father/Mother/Elder Brother, Sunrise magazine*, Pasadena, Theosophical University Press.

Rooke, Andrew, March, 2016, *The Presence of Light*, Theosophy Downunder Newsletter available at http://theosophydownunder.org/australiantsnewslettermarch2016.pdf as on the 13/3/2017.

Rosenblum, Bruce and Kuttner, Fred, 2011, *Quantum Enigma*. Oxford University Press.

Saint Germain on Alchemy: Formulas for Self-Transformation (recorded by Elizabeth Clare Prophet and Mark L. Prophet),

1988, Livingston, Summit University Press.

Schneider, Michael S., 1995, *A Beginner's Guide to Constructing the Universe. The Mathematical Archetypes of Nature, Art, and Science*. Harper.

Sri Aurobindo, 1992, *The synthesis of Yoga.* Silver Lake, Lotus Light Publications.

St. Germain, 1993, *Earth's Birth Changes* (St Germain through Azena). Cairns, Triad Publishers PTY LTD.

Stapp, Henry P., 2011, *Mindful Universe*. Springer.

Talbot, Michael, 2011, *The Holographic Universe.* New York, Harper Perennial.

The Mahatma Letters, 1926, London, T. Fisher Unwin LTD.

The Mysterious Story of X7. Exploring the Spiritual nature of Matter, 2010, Scotland, UK, Findhorn Press.

Trafton Anne, October, 2009, *A New Dimension for Genome Studies,* MIT News Office.

Witten, Edward, 1998, *Duality, Spacetime and Quantum Mechanics*. KITP Public Lectures.

Yogananda, Paramahansa, 1995,*The Bhagavad Gita, Royal Science of God-Realization*. Los Angeles, California, USA, Self-Realization Fellowship.

Glossary

Agni (*Sanskrit*) the God of Fire in *Vedas*; also, literally 'Fire', or spiritual energy of highest frequency vibrations.

Aharata (*Sanskrit*) the energy channel going along the spinal cord.

Akasha (*Sanskrit*) the sphere of pure consciousness, the universal archetype in Spirit's Memory, the Akashic records contain all the past, present and future in their eternal 'now'.

Alchemy a process of transformation of either a human being or matter (particularly a metal) into a pure, noble and high vibrational (more spiritual and powerful) condition.

Archetype the original pattern or model on which all things of the same kind are based, a model or first form.

Astral World a world invisible to most of us, where people can appear in their 'etheric' bodies. Its vibrations may be higher (more of Light) or lower (darker). Our spiritual body appears in the layers of vibrations that conform to our state of consciousness.

Aura radiant human energy, the energy field extending from within to outside the human physical body.

Bija (*Sanskrit*) the original source of the impulse of life to express itself; the source of the impulse of energy animating any creation.

Boddhi (*Sanskrit*) wisdom-knowledge.

Brahman (*Sanskrit*) the impersonal, highest and inconceivable principle which is the origin and final goal of all existing.

Brahma (*Sanskrit*) God in His aspect of Creator; originates from the Sanskrit root 'brih' – 'to expand'.

Buddha (*Sanskrit*) 'the Enlightened' – the highest level of awareness for a human being.

Buddha Sakyamuni (*Sanskrit*) Gautama Buddha – the founder of the teaching of Buddhism. 'Sakya' is the name of his royal kin, and 'Muni' means 'powerful in compassion, solitude and silence'.

Causal Body the highest of the human spiritual bodies assimilating spiritual qualities acquired in the current or former incarnations.

Chakras (*Sanskrit*) energy 'wheels', or revolving centres of the Aharata, providing human organism with energies of different types and qualities.

Coherence a wave property where frequency is constant and corresponding points are in phase. Two wave sources are coherent if they have the same frequency and a constant phase difference.

Cosmic Magnet the cosmic force of attraction towards unification.

Cosmogony science or teaching on the origins of the Cosmos, its spiritual and manifested Worlds, its structure, laws and mechanisms of existence.

Deva-man 'Deva' in *Sanskrit* means 'God' or 'a celestial being' of the Fiery or the Fine World. 'Deva-man' means an entity that is half way between a 'Deva' and a 'man' and possesses the properties of both.

Dharma (*Sanskrit*) universal law, and also spiritual duty fulfilled in any circumstances, in which a man appears.

Domain (Frequency Spectral Domain) – a digital representation of a waveform by a graph of instantaneous amplitude against frequency; frequency – domain graph shows how much of the signal lies within each given frequency band over a range of frequencies.

Elemental Beings (Elementals) primary archetypes, sometimes understood as forces of Nature – spirits of earth, air, fire and water. They are so close to the unity of Spirit they haven't yet developed their individual 'ego' with its conscious understanding of 'self'. That's why they are often perceived as unconscious phenomena.

Entropy lack of pattern or organization; disorder (www.

freedictionary.com/entropy, as on 20/7/2016).

Evolution: 1) development of Matter into Spirit through 'unwrapping' the Spirit from within its forms; 2) unfolding of consciousness resulting in physical development. See also: Involution.

Fiery World the highest of the Three Worlds of existence.

Fine (Subtle) World the intermediate World of existence. See also 'Fiery World' and 'Solid World'.

Fohat (*Sanskrit*) the essence of cosmic energy; the universal power; the super-fine and super-mighty vibrations of primeval energy.

Fractal: 1) geometrical figure, each part of which follows the same pattern as the whole. Similar patterns may recur at progressively smaller scales or in different directions; 2) a mathematical set that exhibits a repeating pattern displayed at every scale (Wikipedia).

Gunas (*Sanskrit*) the essential qualities, or characteristics, of any material substance; the Gunas are three 'threads' forming the fabric of the Nature: Sattwa, Rajas, and Tamas (see under each one of these three).

Heart Chamber the invisible energy body chamber behind the physical heart where the spiritual Fire ignites life in the human spiritual body and initiates heartbeat.

Hologram a manifestation (projection, form) of Thought made by Light waves (vibrations). A form of Matter produced by the Creator's Consciousness. In photography, holography is a method of obtaining three-dimensional photographic images of solid bodies. In the Cosmos, multi-dimensional holograms of universes are manifested as images of the 'perfect pattern' existing in the Creator's Consciousness.The word 'hologram' originates from the Greek 'holus' – 'whole' and 'gram' –'message'. In fact, any part of a hologram manifests the properties of the whole, i.e. carries the message of the whole.

Incarnation embodiment 'in flesh', or one lifetime of the whole lifestream.

Involution the process of differentiation and manifestation (materialization) of Spirit into Matter.

Kalagia: 1)in our understanding, the word 'Kalagia' originates from 'Kal-Agni' ('the Flame of Time' according to H.P.Blavatsky). The book *Kalagia* explains it also as 'power over time' – the stage of spiritual development when consciousness immerses itself into the primordial Fire of the Inner Spirit and gains power over time. "Kalagia!" is also the call to Evolution, 2)Kalagia is the teaching that has been sent to humanity through Alexander Naumkin in the mountains of Altai, in Russia. Naumkin had to climb the sacred mountain of Belucha, and live there 'on the edge' of existence, until he accepted and wrote the book *Kalagia* that was published in Russian in 1993.

Karma (*Sanskrit*) the consequences of action in the material world. Humans are going through their earthly experiences, collecting and balancing Karma. The Law of Karma governs these processes.

Kundalini (*Sanskrit*) the vital feminine 'snake' power, usually latent as a 'coil' of three and a half revolutions in the Muladhara-Chakra at the base of the spine. When Kundalini is awakened and lifted to the crown of the head, the human body becomes purified and enlightened.

Leptons according to modern physics, the basic building blocks of matter, 'elementary particles' including electrons ('charged leptons') and neutrinos ('neutral leptons').

Lifestream the uninterrupted flow of life extending through all incarnations.

Light electromagnetic multi-dimensional wave (radiation) of any possible frequency in the whole frequency spectrum.

Logos (*Greek*) the 'Lord's Word', the first hologram, or manifestation of Spirit.

Mantra (*Sanskrit*) the sound of a prayer, sacred syllable, music or name – vibrations carrying energy, when the sound is uttered.

Mantram (*Sanskrit*) a multi-dimensional environment producing Mantras.

Manu (*Sanskrit*) the manifest Logos, or the spiritual 'Tree of Life' in a planetary chain of manifestation; the multi-dimensional consciousness that comes into manifestation to give birth to any manifest being's consciousness.

Manvantara (*Sanskrit*) the period of Universe's manifestation; literally, 'the time between two Manus'.

Masters, The spiritually advanced Beings who have mastered their earthly experience and liberated themselves from re-incarnation. In some traditions they are understood as living gods and demigods. They usually exist in the higher vibrations of the Fine and Fiery Worlds, and sometimes lower their vibrations by will to manifest in the Solid World. The Masters communicate with us humans as our Teachers, Mahatmas (Great Souls), in different ways, inspiring, guiding and sharing with us their vast Knowledge and Wisdom.

Maya (*Sanskrit*) the Illusion of the temporary manifested world.

Momentum spiritual power that a human being collects and realises as an impulse of energy.

Monad a unit of universal consciousness; indivisible spiritual 'seed' containing the archetype of any entity.

Nadi (*Sanskrit*) energy channels, carrying prana, in the human body, literally 'a river'.

Neith (Net) 'the only Egyptian goddess regarded as 'self-born', 'the mother of Ra' 'who gave birth to the Sun' (see J.S.Gordon, *Esoteric Egypt*, 2015, p. 112). Also known as the goddess of weaving, wisdom and war.

Nonlocality interconnectedness of quantum objects manifesting in their behaviour which the laws of classical physics cannot explain.

Octaves in the spiritual world, as it is in music, all vibrations' frequencies are divided into octaves. The higher octaves of vibrations belong to the higher levels of Existence.

Plasma in physics: one of the four fundamental states of ordinary Matter (solid, liquid, gas, plasma).The examples of plasma are: the matter of stars, lightning and ionized gas. After *Kalagia,* we call 'plasma' 'Matter that is carrying Light', i.e. the substance of any Matter consisting of different frequency electromagnetic vibrations, higher than those of gas. For example, the Matter of Fine and Fiery Worlds up to that of the Prima Materia.

Pralaya (*Sanskrit*) the period of Universe's dissolution.

Prana (*Sanskrit*) the vital fine energy nourishing the human physical body.

Races (Root Races) in theosophy, stages of humanity's development. People of every Root Race possess their particular properties: dimensionality of consciousness, level of spirituality, intellectual power and physical appearance. In *The Secret Doctrine,* H.P.Blavatsky describes the human Races and tells us that the majority of contemporary (at the time when she wrote her books) humanity belongs to the fifth Root Race. There are some representatives of the third and fourth Races remaining on the Earth. The upcoming sixth Race of humanity will be distinguished by higher spirituality balanced with the intellect, and with consciousness reaching the fourth dimension of Being. Early representatives of the sixth Race are now with us and will become more numerous in the future.

Rajas (*Sanskrit*) one of the three aspects of every manifest thing – the three Gunas; the aspect representing action, dynamics, alteration, and passion.

Reactor of Evolution the mighty source of cosmic energy, the main mechanism (structure and process) of the Monad, see *The Laws of Life,* pp 38-39.

Qabalah(Kabbalah) an esoteric method, discipline, and school of thought that originated in Judaism.

Samsara Wheel (*Sanskrit*) the circle of incarnations (lives) a human being is rotating within, collecting and balancing Karma.

Samskaras (*Sanskrit*) the vibratory connections of Karma.

Sattwa (*Sanskrit*) the highest of the three Gunas creating harmony, balance, purity and enlightenment.

Shakti (*Sanskrit*) though there are many meanings of Shakti, in this book it is understood as Universal Energy.

Shambhala the mysterious place somewhere in the Himalayas where the multi-dimensional Essences (the Ascended Masters and their accepted students) are abiding. From this place, the Ascended Masters send their messengers forth into the world at appropriate times. According to St. Germain (*Earth's Birth Changes*), the entire four-dimensional Earth will become Shambhala.

Siva (*Sanskrit*) the God – Destroyer, an aspect of the Hindu Trimurti (Trinity): Brahma, Vishnu and Siva.

Siddhi (*Sanskrit*) manifestation of the divine power in a human being. When one is worthy of the Siddhi on his/her spiritual Path, then he/she is able to apply certain occult powers. This becomes manifested sometimes as gift, talent, and ability of healing and making miracles.

Silver Thread (Sutratma) (*Sanskrit*) the Ray of Light (Spirit's 'working ray') of the hologram that manifests and enlivens the physical human body.

Solid World the lowest of the Three Worlds of existence.

Subtle World see Fine World.

Sushumna (*Sanskrit*) part of the 'Silver Thread', the energy flow along the spine providing energy to the spinal cord and the spinal column of a human body.

Sutratma (Sutratman) (*Sanskrit*) literally, 'thread self' (see 'Silver Thread').

Synthesis the blending of separate elements into the unique whole under a unifying idea.

Tamas (*Sanskrit*) one of the three Gunas, the one representing preservation, inaction, inertia and ignorance.

Theosophy according to H.P.Blavatsky's *The Key to Theosophy*, 'Divine Knowledge or Science', or 'Divine Wisdom', from the Greek 'Theos' – 'God' and 'Sophia' – 'Wisdom'. The term is many thousands of years old; it has been revived by H.P.Blavatsky, who founded the Theosophical Society in 1875 and revealed for the Western readers, in her books, especially in *Isis Unveiled* (1877) and *The Secret Doctrine (1888)*, the esoteric (secret) roots of all the world religions and spiritual teachings. Theosophy proclaims Divine Truth as its foundation, and Universal Brotherhood as its aim.

Thought-form the form created by the vibrations of a thought.

Triunity a multi-aspect esoteric mechanism of creation applied by all spiritual teachings and religions, for example, in Christianity as 'The Father, the Son and the Holy Ghost', or in Hinduism as 'Brahma, Vishnu and Siva'.

Vishnu (*Sanskrit*) the God – Protector and Preserver, an aspect of the Hinduism Trimurti (Trinity): Brahma, Vishnu and Siva.

Index

Angel Healing & Alchemy
How To Begin Melchisadec, Sacred Seven & the Violet Ray
Angela McGerr
Angelic Healing for physical and spiritual harmony.
Paperback: 978-1-78279-742-5 ebook: 978-1-78279-337-3

Colin Wilson's 'Occult Trilogy'
A Guide for Students
Colin Stanley
An essential guide to Colin Wilson's major writings on the occult.
Paperback: 978-1-84694-706-3 ebook: 978-1-84694-679-0

The Heart of the Hereafter
Love Stories from the End of Life
Marcia Brennan
This book can change not only how we view the end of life, but how we view life itself and the many types of love we experience.
Paperback: 978-1-78279-528-5 ebook: 978-1-78279-527-8

Kabbalah Made Easy
Maggy Whitehouse
A down to earth, no-red-strings-attached look at the mystical tradition made famous by the Kabbalah Center.
Paperback: 978-1-84694-544-1 ebook: 978-1-84694-890-9

The Whole Elephant Revealed
Insights Into the Existence and Operation of Universal Laws and the Golden Ratio
Marja de Vries
An exploration of the universal laws which make up the dynamic harmony and balance of the universe.
Paperback: 978-1-78099-042-2 ebook: 978-1-78099-043-9

Readers of ebooks can buy or view any of these bestsellers by clicking on the live link in the title. Most titles are published in paperback and as an ebook. Paperbacks are available in traditional bookshops. Both print and ebook formats are available online.
Find more titles and sign up to our readers' newsletter at http://www.johnhuntpublishing.com/mind-body-spirit
Follow us on Facebook at https://www.facebook.com/OBooks and Twitter at https://twitter.com/obooks